The Social Construction of the Ocean and Modern Taiwan

This book interprets the meanings of the uses, regulations and representations of the ocean undertaken by the state and other societal power sources in modern Taiwan between 1949 and 2016.

Following Michael Mann's historical sociology and Philip Steinberg's political geography, the book analyses the construction of the ocean by the society of Taiwan in terms of ideological, political, military and economic sources of power. It also provides a structural foundation for creating a framework of the politics in maritime and ocean affairs through the lens of an interpretive analysis of the modern Taiwanese construction of the ocean. Moreover, it explores the social constructions of the ocean through the written works of intellectuals in natural sciences, social studies and humanities in Taiwan after the 1980s.

Succinctly revealing how Taiwanese society has influenced the social construction of the ocean, this book will appeal to scholars and students interested in Taiwanese politics and history, political geography and Asian politics.

Kuang-hao Hou is Associate Professor in the Department of Ocean and Border Governance, National Quemoy University, Taiwan, ROC.

Routledge Research on Taiwan Series

The *Routledge Research on Taiwan Series* seeks to publish quality research on all aspects of Taiwan studies. Taking an interdisciplinary approach, the books will cover topics such as politics, economic development, culture, society, anthropology and history.

This new book series will include the best possible scholarship from the social sciences and the humanities and welcomes submissions from established authors in the field as well as from younger authors. In addition to research monographs and edited volumes general works or textbooks with a broader appeal will be considered.

The Series is advised by an international Editorial Board and edited by *Dafydd Fell* of the Centre of Taiwan Studies at the School of Oriental and African Studies.

Series Editor: Dafydd Fell, SOAS, UK

38. Taiwan's Contemporary Indigenous Peoples
Edited by Dafydd Fell, Daniel Davies, Chia-yuan Huang

39. Resounding Taiwan
Musical Reverberations Across a Vibrant Island
Edited by Nancy Guy

40. Taiwan During the First Administration of Tsai Ing-wen
Navigating in Stormy Waters
Edited by Gunter Schubert and Chun-yi Lee

41. The Social Construction of the Ocean and Modern Taiwan
Kuang-hao Hou

For more information about this series, please visit: https://www.routledge.com/Routledge-Research-on-Taiwan-Series/book-series/RRTAIWAN

The Social Construction of the Ocean and Modern Taiwan

Kuang-hao Hou

Routledge
Taylor & Francis Group
LONDON AND NEW YORK

First published 2023
by Routledge
4 Park Square, Milton Park, Abingdon, Oxon OX14 4RN

and by Routledge
605 Third Avenue, New York, NY 10158

Routledge is an imprint of the Taylor & Francis Group, an informa business

British Library Cataloguing-in-Publication Data
A catalogue record for this book is available from the British Library

Library of Congress Cataloging-in-Publication Data
A catalog record has been requested for this book

ISBN: 978-1-032-27098-2 (hbk)
ISBN: 978-1-032-27099-9 (pbk)
ISBN: 978-1-003-29134-3 (ebk)

DOI: 10.4324/9781003291343

Typeset in Bembo
by KnowledgeWorks Global Ltd.

For my mother-in-law,
Nancy Lan-chien Weng

Contents

Acknowledgements

This monograph began as a journal article. I first read Professor Philip E. Steinberg's *The Social Construction of the Ocean* in 2013. In 2014, the *Australian Journal of Maritime and Ocean Affairs* published my paper on the Republic of China's perception of the ocean. The idea of utilising Steinberg's and Professor Michael Mann's theories to write a scholarly volume on the society of modern Taiwan emerged gradually. In 2015, I met my wife, Jennifer Chieh-fei Li, who supported my research wholeheartedly. We got married in July 2017; I completed the third chapter of this monograph just before our wedding. The notion of speeding up my writing and our baby were both conceived on our honeymoon. The arrival of my wife and daughter was personally fortunate for me, and this monograph has benefitted from my mother-in-law's benevolence.

My mother-in-law, Nancy Lan-chien Weng, kindly took (and still takes) great care of my daughter. I firmly believe this book would never have been completed without her compassionate assistance. I would thus like to dedicate this monograph to my mother-in-law.

I am also extremely grateful to my sister-in-law, Gwen Chuan-chuan Fan, for looking after my daughter and offering us advice on child-rearing. My aunt, Trissy Chunqing Qiu, has also generously offered a helping hand when we needed it. I really appreciate her babysitting my daughter. I am, moreover, thankful for my daughter's nanny, Meiyin Jie, and her assistance during the daylight hours of every weekday. Furthermore, my father-in-law, Professor Nigel Nien-tsu Li, has never stopped caring for my writing; he keeps encouraging me to conduct further research, and I am truly obliged to him.

Professor Lawrence Kuen-chen Fu first invited me to write about marine affairs when I joined the Graduate Institute of Marine Affairs at National Quemoy University in August 2011. Professor Peter Kien-hong Yu was sympathetic towards my research on the maritime and ocean affairs of modern Taiwan. He graciously agreed to read my work and commented constructively on the first case study of this monograph. Professor Dachi Liao also acknowledged the viability of my study and endorsed the value of this monograph. She enhanced my confidence in undertaking such an academic pursuit. Professor Chen-kuo Hsu confirmed the utility of Mann's theory for this book, and Professor Hao-yu Wang encouraged my writing, highlighting the importance of monographs for academia. I am deeply indebted to my teachers, particularly those mentioned earlier.

My mother tongue is Chinese, and my English writing skills require a copy-editing service to conform to conventional English. I am ecstatic with the services provided by Enago, which are funded by the Ministry of Science and Technology of the Republic of China (MOST 110-2410-H-507-001-). I am also grateful for the support extended by the two anonymous reviewers of this project.

This monograph could only be published because of the generous encouragement and guidance I received from Professor Dafydd Fell and Ms Stephanie Rogers. The title of this monograph is kindly advised by Ms Rogers. Mr Andrew Leach also answered my questions considerately and offered his help unstintingly. The professional copyediting and typesetting process managed by Ms Nancy Rebecca is highly appreciated. I wholeheartedly appreciate their assistance, patience and enthusiasm for outstanding and superior quality publications.

Above all, as always, I am solely and fully responsible for this monograph. Any mistakes that may be observed in this academic pursuit are mine to own.

Kuang-hao Patrick Hou
National Quemoy University

Note on Romanisation

Generally, the Romanisation of the names belonging to people and places in Taiwan follows the Wade-Giles system. However, there are at least two problems with employing this system for academic monographs. First, the Chinese characters used in people's names are pronounced differently depending on the Chinese dialect used, and people may choose to Romanise their names according to their dialects. For example, SUN Yat-sen is the Romanisation when Cantonese (or Guangdongese) is used rather than Mandarin. It is very difficult, if not impossible, to know how one chooses to spell their Chinese name using the Wade-Giles system without checking their passport. Second, even when one uses Mandarin to pronounce their name, the Wade-Giles system does not accurately match the pronunciation. For example, my name is HOU Guanghao in Mandarin, but my name is spelt HOU Kuang-hao on my passport because the Ministry of Foreign Affairs adopts the aforementioned system, transforming the sound of 'G' to 'K'. Moreover, a Chinese character with a single pronunciation can be Romanised into different spellings by the system. For example, the surname Li can sometimes be spelt as Lee, as exemplified in the name of a prestigious law firm in Taiwan, Lee and Li Attorneys-at-law.

Therefore, in this monograph, I adopt *Hanyu Pinyin* for the Romanisation of persons' names mentioned in the main texts. For example, JIANG Jieshi is used instead of CHIANG Kai-shek (which is the Cantonese pronunciation). The Romanisation of authors and academics' Chinese names discussed and referred to in this monograph follow their own preferences, given that the spellings of their names are frequently provided in their works. It is very easy to verify the spelling, and there is no need to check their passports. The Chinese names of places follow their official Romanisation—for example, Keelung and Kaohsiung. More importantly, I believe it is easier and less confusing for English language readers to know whom I am referring to in the main texts if I use *pinyin* because this system is more commonly used around the world than the Wade-Giles convention.

List of Abbreviations

APOC	Asia-Pacific Operation Center
CCP	Chinese Communist Party
CIA	Central Intelligence Agency
CMSN	China Merchant Steamship Navigation
CMT	Chinese Maritime Transport
COSCO	China Ocean Shipping Company
CSB	CHEN Shuibian
DPP	Democratic Progressive Party
DWT	Deadweight Tonnage
ECFA	Economic Cooperation Framework Agreement
EEZ	Exclusive Economic Zone
EM	Evergreen Marine
FIC	First Island Chain
GDP	Gross Domestic Product
ICCAT	International Commission for the Conservation of Atlantic Tunas
ICFA	International Coalition of Fisheries Association
IUU	Illegal, Unreported, and Unregulated
JCRR	Joint Commission on Rural Reconstruction
JJG	JIANG Jingguo
JJS	JIANG Jieshi
KMT	Kuomintang
LDH	LI Denghui
MOT	Ministry of Transportation
MYJ	MA Yingjiu
NAD	Navy Advisory Division
NTD	New Taiwan Dollar
PLA	People's Liberation Army
PLAAF	People's Liberation Army Air Force
PLAN	People's Liberation Army Navy
PRC	People's Republic of China
ROC	Republic of China
ROCA	Republic of China Army

ROCAF	Republic of China Air Force
ROCN	Republic of China Navy
SAMDT	Sino-American Mutual Defense Treaty
SLOC	Sea Lanes of Communication
TEU	Twenty-foot Equivalent Unit
TIM	Taiwan Independence Movement
TN	Taiwan Navigation
TPP	Three Principles of the People
TRA	Taiwan Relations Act
TSU	Taiwan Solidarity Union
UK	United Kingdom
UNCLOS	United Nations Convention on the Law of the Sea
UN	United Nations
US	United States of America
WHL	Wan Hai Lines
WTO	World Trade Organization
YMMT	Yang Ming Marine Transport

1 Introduction

Ocean, Politics and the Society of Taiwan

Introduction

This monograph introduces the social constructions of the ocean as formulated by four important sources (that is, ideological, military, economic and state sources) of social power in the modern society of Taiwan between 1949 and 2016.[1] This proclamation of the intention of this study contains abstract concepts and theories that require further explication, which is the task of this chapter. The next section addresses the overarching question that this monograph aims to answer. In addition, an analytical framework is proposed that structures this monograph and provides meanings for the phenomena being explored. The third section elaborates on the social context through which the topic can be better understood. The final section provides the layout of this monograph.

The Overarching Research Question and the Analytical Framework

The Overarching Research Question

The overarching research question is simple but fundamental: how did the state and other power sources of society of modern Taiwan mainly construct the ocean between 1949 and 2016?

The state currently effectively governing Taiwan (including Kinmen and Matsu; prior to 1949, the whole Chinese mainland), formally called the Republic of China (ROC). It promulgated two marine-territorial laws in 1998, namely, the Law on the Territorial Sea and the Contiguous Zone of the Republic of China and the Law on the Exclusive Economic Zone and the Continental Shelf of the Republic of China.[2] Afterwards, the ROC formulated a series of marine policies, which can be found in a series of white papers. After 2001, these white papers include *Ocean White Paper* (Research, Development, and Evaluation Commission 2001), *National Oceans Policy Guidelines* (Coast Guard Administration 2004), *Ocean Policy White Paper* (Council of Marine Affairs Advancement 2006) and *Ocean Education Policy White Paper* (Ministry of Education 2007).

DOI: 10.4324/9781003291343-1

The following problems do not comprise the main research question to be addressed. Instead, the main research question can be derived from these problems. How were these policies implemented? Were significant difficulties encountered during their implementation? Lastly, what efforts did the state make to resolve such difficulties? To paraphrase these questions with a strong accent of political science, one may ask: how did other societal power sources render the implementation of the policies of the state increasingly difficult? What can the state do to overcome the oppositions or incorporate the interests of other societal power sources? This monograph will elaborate on the definition of the notion of power sources. In the meantime, one may also ask: Why did the state need to do so?

The examples of the difficulties faced by the state in its attempt to implement marine policies include the stipulation of the Coastal Zone Management Act and the temporary suspension of the establishment of the Ocean Affairs Council. The Ministry of the Interior first drafted the Coastal Zone Management Act in 1991. The bill of this act had been delivered to the Legislative Yuan (that is, the parliament of the ROC) for deliberation four times without fruitful consequences before it was finally passed in January 2015.

A similar difficulty faced the establishment of the Ocean Affairs Council. The state controlled by the Kuomintang (KMT, which is also known as the Chinese Nationalist Party) stipulated the Organizational Act of the Executive Yuan in 2010 and decided to form the council. In 2015, the then President MA Yingjiu (also known as MA Ying-jeou; hereafter, MYJ), promulgated the act. However, the inauguration of the council was suddenly suspended in July 2016 after the Democratic Progressive Party (DPP) took over the presidency in May. As a result, the inauguration of the council was postponed to April 2018.

The abovementioned examples are well-known, at least amongst scholars of marine affairs and the concerned state officials of the ROC. These examples demonstrate that the aforementioned questions about the politics of the stipulation and implementation of marine policy in Taiwan are worth exploring. However, as previously implied, this monograph does not embark on such an exploration. It analyses a fundamental question, whose answer can reveal the socio-structural context of the stipulation and implementation of marine policy, which will, thus, enable scholars of marine policy to generate a comprehensive picture of and deep insight into problems similar to those mentioned earlier.

Effectively, the aforementioned problems are based on a presumption, that is, the state and other power sources in the society of modern Taiwan between 1949 and 2016 utilised the ocean in a different manner. Specifically, the functions served by the ocean and the meanings represented by the ocean for the state and other societal power sources presumably differed from one another. The results of such differences are conflicts of interests in utilising the ocean and coastal areas.

Nevertheless, this monograph demonstrates the abovementioned presumption through the lens of territorial political economy purported by Steinberg (2001) and his classification of the social constructions of the ocean. Hereafter, the Steinbergist Classification for brevity. After presenting the presumption, this monograph paves the path to answering the series of problems addressed earlier for future research. Although this monograph does not provide concrete answers for these problems, they can be addressed by other projects.

In summary, this monograph derives the overarching research question from the aforementioned presumption. Critically, this monograph intends to answer the following question: how did the state and other societal power sources construct the ocean during modern Taiwan between 1949 and 2016? The answer constitutes an elucidation of the three main theses that underlie Steinberg's (2001) theory of territorial political economy. First, I demonstrate that the use, regulation and representation of the ocean mutually influence one another and, in turn, shape the social construction of the ocean. Chapters 2 and 3 present this demonstration.

Second, I illustrate the view that the geographical condition and the material organisation of society can shape the social construction of the ocean. Chapter 4 provides an illustration of this perspective. Third, I exemplify the effect of the social construction of the ocean on the material organisation of society and on the first nature of the ocean, as coined by Steinberg (2001). Chapter 5 presents this exemplification.

Effectively, by explicating the social constructions of the ocean on modern Taiwan, this monograph not only elucidates Steinberg's (2001) theses on the theory of territorial political economy but also points out, examines and synthesises important issues in the maritime and ocean affairs of modern Taiwan into a meaningful account. Moreover, this monograph can set a foundation for the analysis of the politics of the stipulation and implementation of marine policy.

At this point, a clarification of the concepts and theories involved in the primary question is required. Effectively, such concepts and theories constitute the chief analytical framework of this monograph.

The Analytical Framework

The Theory of Territorial Political Economy

The first step to understand the overarching research question is to grasp the theory of territorial political economy. As Steinberg (2001) points out, this theory is useful for analysing not only the ocean but also the land and other geographical phenomena. This theory has been influenced by several theoretical schools, such as human geography, cultural ecology, Marxist political geography and the theory of the social production of space. However, this monograph does not intend to introduce these theoretical paradigms in detail. Nevertheless, their contribution to Steinberg's theory of territorial political economy can be summarised as follows.

First, human geography points out the necessity of examining the intersections and interactions amongst 'the social structures, individual behaviours, institutional arrangements, and natural features' (Steinberg 2001: 20). Furthermore, human geography highlights that natural features become the limiting or enabling factors of certain human actions.

Second, cultural ecology criticises human geography for presenting the human-nature relationship as a single-directional oversimplification. It indicates that the flows of influences between humans and nature are bi-directional. As Steinberg (2001) mentions, 'Human actions shape the physical landscape, just as the landscape enables and limits human action and social formation' (Steinberg 2001: 20).

Third, Marxist political geography refines the abovementioned notion of cultural ecology and adds that the endeavours of a person or group to change or transform nature can effectively lead to the emergence or alteration in the institutions of a society (Steinberg 2001).

Finally, theorists of the social production of space amend the Marxist political geography by arguing that humans intrinsically and inevitably influence and are influenced by the constantly evolving natural space. This continuous evolution of space is a result of the involvement of humans in the social production of space.

Based on these theoretical predecessors, the theory of territorial political economy consists of three major components. Steinberg (2001) states that these three parts are nature, material organisation of society and social constructions of space. By nature, it is straightforward and indicates geographical phenomena, such as a mountain range. For humans, the ocean is one of the most important geographical phenomena, where its physical conditions or 'first nature' continues to be explored by humans.

The material organisation of society refers to social practices, rules, organisations and technologies that are conducted, stipulated, formed and employed to transform and exploit geographical phenomena or nature. For example, the modern capitalist system and the early modern mercantilist system are two material organisations of society (Steinberg 2001).

The social construction of space or the second nature of a geographical phenomenon denotes the ideational and behavioural complex created by human use, regulation and representation (that is, interactive elements) of the geographical phenomenon. These elements can be enabled or constrained by the physical conditions of a geographical phenomenon (or nature). Moreover, they can interact and influence one another and, thus, create the second nature of the phenomenon. In other words, the social construction of nature is generated through the differentiated agglomeration and fusion of these three interactive elements of nature (Steinberg 2001). Chapters 2 and 3 demonstrate this point.

These elements of the social construction of nature can be further explicated. To begin, Steinberg (2001: 32) defines representation as a process 'by which social meaning is transmitted among individuals through literary creations, visual images, and other media'. Literary creations, visual images and other media are discourses. However, Steinberg (2001) focuses on three types of discourses in the modern era, namely, developmental, geopolitical and legal, to formulate his theory of territorial political economy due to the abundant number of discourses.

In developmental discourses, each geographical phenomenon on land can be 'represented as a canvas on which planners and engineers may test and apply their insights and work toward human progress' (Steinberg 2001: 34). In these discourses, the ocean is represented as an untameable void that resists rational planning, development and progress.

Meanwhile, geopolitical discourses state that, 'the key unit is the territorially defined state that interacts with the world's other states' (Steinberg 2001: 34). In these discourses, the ocean is represented as a surface on and across which states can wield military power to dominate other states and competitors. In other words, such discourses represent the land as territorial states separated by the ocean as an empty arena.

In legal discourses, social relations are typically operationalised and legalised. Therefore, the legal discourses on certain geographical phenomena are inclined to define and demarcate places, to arrange their hierarchical order and to identify the scopes and boundaries of social organisations. Regarding the ocean, 'the legal discourses imply that the sea is "lawless", an antithetical "other" lying outside the rational organization of the world' (Steinberg 2001: 35).

The abovementioned forms of discourses represent the ocean as an empty space, which lacks the potential of development, is unclaimable as a state territory or is uncontrolled and without order. Nevertheless, Steinberg (2001) extracts the two other elements (that is, use and regulation) of the social construction of nature from his study on the three types of discourse.

According to Steinberg (2001), one can understand the hierarchical rank of relevance that humans assign to a certain geographical phenomenon and the approach adopted to derive functions from it by evaluating its use. Regarding the use of the ocean, the ocean may be vital for several people that use it or may be rendered with a marginal role in obtaining wealth.

By examining the regulation of a state over a certain geographical phenomenon, one can comprehend whether the state territorialises it. The regulation of the ocean is mainly concerned with the levels of state intervention, domination, governance and possession. Many civilisations may deem the ocean as a property, whereas others may merely view it as a dangerous void.

Having introduced the three elements of the social constructions of geographical phenomena, this monograph can now summarise the relationships amongst the three elements of Steinberg's (2001) theory of territorial political economy.

First, the use, regulation and representation of a person or a group of the geographical phenomenon are closely linked to established social practices, institutions and organisations, as well as technologies employed to transform and exploit the geographical phenomenon. In other words, the social construction of the geographical phenomenon cannot be isolated from its wider context, that is, the material organisation of society. Second, the material organisation of society reflects the conditions of nature. Simply put, the physical conditions of the geographical phenomenon constitute the natural environmental context for the material organisation of society (Steinberg 2001). Chapter 4 illustrates these two points.

Third, the flow of influences can be inverted. The aforementioned first and second points illustrate that the dynamics of influences begins with nature and moves towards the material organisation of society, which then pushes forward to social construction. In the same manner, this flow of influences can move in the opposite direction. The newly emerged social construction of the geographical phenomenon can influence the establishment of organisations and institutions and the employment of technologies to exploit and transform the physical conditions of the phenomenon. In turn, these aspects influence and change the physical conditions (Steinberg 2001). Chapter 5 discusses this point.

The Steinbergist Classification

I have briefly elaborated the essence of Steinberg's (2001) theory of territorial political economy. Effectively, this monograph can help readers better understand this topic. Based on the theory, he also develops a classification of the social constructions of the ocean, that is, the Steinbergist Classification, which consists of three primary models, namely, the Indian Ocean Model, the Micronesia Model and the Mediterranean Sea Model. The Classification was constructed with historical evidence collected from various places: 'the Indian Ocean prior to large-scale European influence, Micronesia prior to very recent times and the Mediterranean under Roman rule' (Steinberg 2001: 6).

To formulate the Indian Ocean Model, historical evidence was extracted from roughly between 500 BC and AD 1500 and in the geographical area spanning from 'the east coast of Africa to the islands of Indonesia to the southern coast of China' (Steinberg 2001: 41). Historically, long before the arrival of Vasco da Gama in 1498, extensive sea links were present in this area. The sea provided access to communication among different coastal civilisations. Notably, the Chinese merchants dominated traded in the eastern Indian Ocean between the eighth and fifteenth centuries, whereas the Arabs dominated the western Indian Ocean (Steinberg 2001).

However, the physical conditions of the seas in this area compromised the full integration of these coastal civilisations through sea routes. Moreover, seasonal monsoons constrained naval and maritime activities. Thus, undertaking more than one round trip per year on long-distance sailing routes was impractical. As such, overseas military occupation and frequent maritime trade in bulk commodities were obstructed and limited. The concentration of wealth lay inland along rivers through which a large agricultural surplus could be easily accessed (Steinberg 2001).

'While the sea was a presence in Indian Ocean social life, it was constructed as a special space of trade, external to society and social processes' (Steinberg 2001: 45). Thus, the ocean was constructed not only as a source of foreign goods, but also as an untameable, dangerous void (Steinberg 2001). Additionally, many Chinese people from the southern coastal provinces, such as Fujian and Guangdong, opted to leave their home country through sea routes. The land on the other side of the ocean, but not the ocean itself, was perceived as a chance for a better life. The sea was a mysterious surface and void across which trade and migration could be conducted.

Furthermore, following the aforementioned notion of the ocean as existing only as a surface, the ocean was considered neither suitable for state intervention and governance nor possessable as a territory. 'The sea was perceived not as territory but as a space of distance and the duty of the merchant sailor was to annihilate that distance to the best of his abilities' (Steinberg 2001: 46). Additionally, the states of the coastal civilisations surrounding the Indian Ocean perceived the lack of benefits from establishing and maintaining strong naval forces or from colonising overseas lands. The connections between these coastal civilisations were mainly economic and cultural instead of military and political (Steinberg 2001).

In summary, the Indian Ocean Model views the ocean as similar to a desert. The ocean has been constructed as a shapeless and ungovernable, undifferentiated void that should be crossed to conduct terrestrial activities.

The Micronesia Model was built with evidence derived from the societies of Micronesia. 'The societies of Micronesia, unlike those of the Indian Ocean, traditionally have constructed the ocean as territory' (Steinberg 2001: 52). Moreover, the seas were constructed not as an empty space of distance but as a reservoir of resources. Therefore, the ocean was to the Micronesians as the land was to continental residents (Steinberg 2001).

A fundamental character of the Micronesian concept of the ocean was that it was rife with a sense of places. The following example given by Steinberg (2001: 54) is worth quoting. A Micronesian navigator …

> [m]ight, for instance, know that his destination island lies under Star A. Because of the specific ocean currents prevailing at this time of the year, he should aim for Star B. However, if when Star C is at a certain point in the sky, he sees a reef on his right and he feels a leeward ocean current on his boat, he will know that he has drifted off course and that he should aim for Star D to compensate. But as long as he is over the reef, he may want to drop a line, because this 'rest area' is known to be particularly abundant in a prized variety of fish.

For Micronesians, the ocean consisted of various adjacent marine areas as distinct places. Islands, reefs or rocks became special characteristics of such places. They are similar to 'significant "road signs" of the highways of the sea' (Steinberg 2001: 55). In addition, the ocean became a highway *qua* place that connects terrestrial and marine places of destinations. 'For the Micronesians, land appears at the end of place-specific water-routes' (Steinberg 2001: 56). As such, oceans are not barriers to be annihilated.

Constructing the ocean as a place has been closely related to its construction as a resource reservoir. In Micronesia, land was small and barren. Although the land provided Micronesians with protection from elements and a place to rest in comfort, the major activities were concentrated on the sea. Where the sea offered food, other necessities and a medium of transport, the ocean connected places and provided resources (Steinberg 2001).

For the Micronesians, the ocean also meant possessions and could be territorialised by the state. 'There is no Micronesian concept of a high seas zone beyond possession. The area of ocean-space "owned" by one island extends until it borders the ocean-space of another island' (Steinberg 2001: 58). Nevertheless, for Micronesians, possessing a marine area was dependent on the function it provides. However, possession itself was not the goal. Claiming a marine area territory was equivalent to exploiting the resources that the area could provide (Steinberg 2001).

In conclusion, the Micronesia Model views 'the deep sea as territory. The ocean is seen as a resource-providing space, characterised by numerous specific places' (Steinberg 2001: 59). Moreover, this Model perceives the ocean as an integrated part of society. People who construct the ocean from the perspective of the Micronesia Model 'govern the sea much as they govern land: as a set of discrete places to be demarcated and controlled according to the general organizing principles of territoriality' (Steinberg 2001: 60).

The Mediterranean Sea Model was created mainly on the basis of the historical experiences of the Roman Empire. It has three variations; however, the essence of the original Mediterranean Sea Model should be first discussed. In the Roman and earlier eras, 'the peoples of the Mediterranean constructed the sea as a non-possessable space, but one in which and across which, state power legitimately could be asserted in the interest of stewarding its bounty' (Steinberg 2001: 61). For the Romans, although the ocean was not a territory, it remained subject to military domination.

Steinberg (2001) identifies that, although the Romans claimed military domination over the Mediterranean Sea, they did not claim the right to own, possess and dispose of the ocean. For them, the seas were a placeless surface that belongs to no one. However, states could wield military forces to control it to ensure the smooth transportation of supplies and troop movement. The Roman military intervention over the ocean focused on the removal of military challenges for their domination over the ocean space and possession of neighbouring lands (Steinberg 2001).

Lastly, this model posits that the ocean is constructed as a space. However, in contrast to the Indian Ocean Model, which constructs the ocean as a space of passages (for trade and migration), the original Mediterranean Sea Model constructs the ocean as a space for arenas (for military competition and domination). Furthermore, contrary to the Micronesia Model, which constructs the ocean as a place for possession, the Mediterranean Sea Model argues that the ocean is not possessable. Instead, this model puts forward that the ocean is constructed as a spatial arena in the sense that states compete with one another to command and control sea lanes but not to possess the ocean itself (Steinberg 2001).

Furthermore, as derived from the academic-political debates amongst Seraphim de Freitas, Hugo Grotius and John Selden, Steinberg (2001) proposes three variations of the Mediterranean Sea Model, namely, Freitian, Grotian and Seldenian. Among the three, the Freitian Variation largely resembles the original Mediterranean Sea Model.

Conceived in the era of mercantilism, the Freitian Variation emphasises the necessity of a state to govern the sea routes that undertake the trade and transport of the state. This variation differs from the original Mediterranean Sea Model in terms of the emphasis on seaborne transport for trade and the need for the state to govern maritime routes. However, the original version stresses military domination over sea areas and lanes. In particular, the Freitian Variation is focused on merchant instead of naval activities on the surface of the ocean. Nevertheless, this variation does not construct the ocean as a possessable territory; thus, the seas as space cannot be owned (Steinberg 2001).

The Grotian Variation constructs the ocean as space. However, it is subject to natural law instead of sovereign will. Moreover, the ocean cannot be possessed by any single state. However, this variation suggests that the community of all states in the world should steward the ocean. This variation deems navigation and fish as inexhaustible resources and views the ocean as a space for spatial arenas on which states can pursue their interests. Nevertheless, its provision indicates that each state should not interfere with the access of other users to navigation and fisheries (Steinberg 2001).

The Seldenian Variation presents a construction of the ocean that differs from those of the two abovementioned variations. Coastal waters and fishery locations in the sea are constructed as territories of pertinent coastal states, which should be governed for the interests of the states. Furthermore, sailing routes can be controlled and governed as required by the states. The ocean beyond the coastal waters and fishery locations, however, is constructed as an empty space that cannot be possessed. Thus, this variation continues to differ from the Micronesia Model because it views the entire ocean as possessable places (Steinberg 2001).

Additionally, the Seldenian Variation, in concurrence with the Grotian Variation, admits the principle that a state cannot legitimately prohibit other states from navigating its coastal sea areas, although it maintains the right of appropriation and possession of the coastal waters and fishery locations. Moreover, this variation proposes that a state's claim of the ownership of a portion of the ocean is feasible given that it can exert effective control. Thus, it matches the Mediterranean Sea Model, because this variation continues to view the control and governance of sea routes as necessary for the state and constructs the ocean as a space for arenas (Steinberg 2001).

The IMES Model

To answer the overarching research question of this monograph, it also employs the IMES (Ideology, Military, Economy and State) Model by Michael Mann (2012a; 2012b). The IMES Model, together with the Steinbergist Classification and the theory of territorial political economy, constitute the primary analytical framework of this monograph. The IMES Model is useful for this monograph because it defines society and identifies its important constituents. However, prior to elaborating on the IMES Model, I would like to discuss the importance of employing analytical frameworks in social studies. In addition, the reason elucidates Mann's (2012a) distinct definition of society.

First, an analytical framework can be a simple concept or can consist of many related concepts and theories. An analytical framework is a product of the conceptualisation and theorisation of a certain phenomenon. It helps researchers realise the focus of their interest and generates meanings for the concepts being examined. Conceptualisation and theorisation are not activities separated 'from the analysis of empirical data. Analyses can only be made in terms of theoretical schema and propositions' (Wallerstein 1974: 347). Moreover, concepts and theories resemble 'a large-scale map of an extended terrain, such as an airplane pilot might use in crossing a continent' (Moore 1966: xiv). In addition, 'all research involves some [concepts and theories], so the question is less *whether* you should use theory than *how* you should use it' (Neuman 2003: 42, original emphasis).

A concept is composed of a label and a definition. It typically refers to a phenomenon, and its definition is an arbitrary, abstract derivation and description of the characteristics and attributes of the phenomenon. Furthermore, a theory is defined as 'a system of interconnected abstractions or ideas that condenses and organizes knowledge about the social world' (Neuman 2003: 42). However, what aspect renders the use of concepts, theories and analytical frameworks inevitable to social studies?

Social reality is 'an infinite flux of events all equally meaningful or meaningless' (Abrams 1982: 77). In other words, daily experiences do not form consistent and complete knowledge about reality. Facts and the social world in which humans live are incompletely and selectively perceived (Parsons 1949). However, the selected perception is not undertaken according to any objectively given criterion. The targets of perceptions 'are selected and constructed out of the infinity of phenomena by being first defined as meaningful within our sense of cultural significance' (Abrams 1982: 78).

In investigating the sources of social power, Mann (2012a) concurs with the aforementioned view and elevates it by indicating that society is, in reality, more complex than sociological theories. Regardless of whether scholars are searching for single- or multi-cause factors, social reality is distorted in their theories, although such a search is an inevitable endeavour to acquire knowledge about social phenomena.

Corresponding to the complex social reality, the IMES Model was designed to address the disarray in society and disordered social change. It is an analytical framework that imitates and represents chaotic social practices and society *per se*. Furthermore, it aims to reflect social reality as much as possible and begins with a definition of society as being constituted of multiple overlapping and intersecting sources of social power.[3] To understand Mann's (2012a) definition, I begin by presenting a definition of social power (or, simply, power).

For Max Weber (1978), social power is an opportunity to impose one's will on the other or others regardless of possible resistance. However, this definition of power conveys the concept that any power relation is merely a zero-sum game. This view can constrain analyses because the only possible examination of power defined is that by scrutinising behaviours in confrontational relationships. Thus, the object of the examination can then only be 'an observable *conflict* of (subjective) *interests*, seen as express policy preferences, revealed by political participation' (Lukes 2005: 19, original emphasis). Thus, power relations become merely confrontational relations (Hou 2012).

This view neglects the collective aspect of power. Humans in co-operation can reinforce their joint capacity to influence nature or a third party (Mann 2012a). For example, in a head-on military clash in ancient times, an army that utilises a well-coordinated formation may generate increased collective crushing momentum to disperse a military band without proper organisation. Thus, this monograph adopts a broad definition of power by viewing power as a transformative capacity (Giddens 1987; Mann 2012a).

Power defined as transformative capacity denotes 'the capability to intervene in a given set of events so as in some way to alter them'. Furthermore, power as a transformative capacity 'must be related to the resources that actors employ in the course of their activities to accomplish whatever they do' (Giddens 1987: 7). In other words, power is exercised via resources (Mann 2012a). For example, the knowledge of conducting a modern war can be viewed as a resource possessed by professional military officers. As such, their knowledge of modern warfare enables them to command an army (Hou 2012).

Resources refer to 'any things, both material and non-material, that people use to further their own desired ends, as individuals or collectively in groups' (Leftwich 2004b: 106, original emphasis). Resources are typically scarce and unevenly distributed. In general, participants in interactions possess various levels of resources and, thus, power. Although this scenario may lead to asymmetrical relations between the participants involved, it does not diminish the possibility of co-operation. In the abovementioned example, commanding military officers and army soldiers can co-operate and defeat their enemy. In other words, the uneven distribution of resources does not necessitate zero-sum relationships; they are dependent on institutions (Giddens 1987).

Power is also exerted through institutions (Giddens 1987). From the collective perspective of power, generating joint capacity requires organisational and institutional devices and techniques. A factory that produces cars can illustrate this point. Factory employees that work according to certain standard operating procedures can increase collective power and produce more vehicles than does each unorganised individual labourer (Hou 2012).

Notably, from the conflicting, distributive perspective, if such interactions are reproduced over a period of time, these interactions within the asymmetrical possession of resources between actors can be institutionalised. The positions and structures of an institution confer tasks to the actors of the institution. In turn, these tasks may generate functions (and, unintentionally, malfunctions); however, an actor undertaking the task of a certain position produces collective power for the entire institution. At the same time, this actor yields distributive power amongst the other actors of the institution (Hou 2012).

For example, one employee is appointed as a factory supervisor, whereas another is a subordinate factory worker. Given that other conditions remain equal, the former holds more power (at least formal, institutionalised power) than the latter. Moreover, one can argue that power as a transformative capacity is generated by and exercised through distributed resources and constructed institutions (Giddens 1987; Hou 2012).

Domination, as a concept closely related to power, can also be defined, because it can help to better understand the original Mediterranean Sea Model. Social interactions of power are also conducted between the autonomous and the dependent because interactions necessarily involve resources and institutions. When these resources are unequally distributed and constructed, the autonomous and the dependent emerge (Giddens 1987). If one observes a situation in which a postgraduate student is discussing a dissertation with a supervisor, then the supervisor becomes autonomous, whereas the student becomes dependent on the supervisor's instructions or permission to conduct a particular research. Thus, a form of domination emerges (Hou 2012). Therefore, domination is a structured asymmetry of constructed institutions and distributed resources between the autonomous and the dependent (Giddens 1987).

Moreover, Mann (2012a) offers three sets of classification of power, namely, distributive and collective power; extensive and intensive power; and authoritative and diffused power. In the first set, distributive power 'restricts its meaning to mastery exercised over other people' (Mann 2012a: 6). For example, an army captain holds distributive power over his soldiers. In contrast, collective power refers to enhanced joint power wielded by persons in co-operation over nature or third parties. A traditional example of collective power may be represented by organised labourers who built the Grand Canal in Imperial China. In the classification, where extensive power 'refers to the ability to organize large numbers of people over far-flung territories in order to engage in minimally stable cooperation', intensive power 'refers to the ability to organize tightly and command a high level of mobilization or commitment from the participants, whether the area and numbers covered are great or small' (Mann 2012a: 7). Lastly, authoritative power 'comprises definite commands and conscious obedience', whereas diffused power 'spreads in a more spontaneous, unconscious, decentered way throughout a population, resulting in similar social practices that embody power relations but are not explicitly commanded' (Mann 2012a: 8). The last two sets can be combined to present four 'ideal-typical forms of organizational reach' (Mann 2012a: 8). The examples given by Mann (2012a) include an army command structure for authoritative and intensive power; a general strike for diffused and intensive power; an ancient military empire for authoritative and extensive power; and market exchange for diffused and extensive power. Theoretically, the strongest power source is one that can combine the four forms.

To enhance the understanding of the abovementioned definition of power, the subsequent text briefly discusses the concept of the institution, which is defined in light of neo-institutionalism. In this tradition, an institution refers to formal and informal rules, which structure actions. Institutions assist humans in achieving their goals in social interactions. B. Guy Peters (1999) proposes that the institution is composed of four core features. First, an institution can be viewed as a structural feature of society, that is, formal (such as the voting procedure in a national assembly) and informal (such as a set of shared norms of a class of students). Peters (1999: 18) further explained as follows: 'As such, an institution transcends individuals to involve groups of individuals in some set of patterned interactions'.

To further elaborate on this characteristic, the definition of the institution can be viewed as rules for games (North 1990).

> They are pervasive in all human societies, shaping social, economic, and political behaviour, from marriage and most aspects of etiquette, to rules of the road, patterns of property rights, ownership and exchange, and modes of representation and action in politics.
>
> (Leftwich 2009: 7)

Moreover, distinguishing the concept of the organisation from that of the institution is important and necessary.

According to Adrian Leftwich (2009), the organisation is a subtype of the institution. If institutions represent the rules of the games, then organisations are the players. The organisation is characterised by three features, namely, possessing 'their own internal rules which apply only to their members' (Leftwich 2009: 7); having 'relative autonomy from each other' (Leftwich 2009: 8); and having 'its own internal structure of power and authority' (Leftwich 2009: 8). Examples of the organisation include military divisions, universities, commercial companies and government ministries.

The second feature of the institution is that it can obtain a certain degree of stability (Peters 1999). For example, a president may convene a coffee meeting in her kitchen every Monday morning to discuss economic policies during her time in office. Thus, this meeting becomes an institution. The third feature is that the institution is important in that it influences human actions or, at least, constrains its members (Peters 1999). For example, the regulation of teachers' promotion can influence their fulfilment of duties. The fourth feature of the institution is that 'there should be some sense of shared values and meanings among the members of the institution' (Peters 1999: 18). For example, in a seminar, the tutor and students share the norm that the tutor should lead the discussions.

Thus far, I have defined social power and institutions. They are the conceptual basis for understanding Mann's (2012a) definition of society as overlapping and intersecting sources of social power. The next step was to explicate his definition of society by introducing these sources. Mann (1988) identifies four extremely important sources of social power: ideological, military, economic and state sources of power.

First, the resources of ideological power sources pertain to concepts and theories (which humans require to interpret and explain the universe and phenomena); norms and values (which humans rely on for social cohesion and appropriate conduct); and religions and aestheticism (which enables humans to express emotions, to apprehend meanings in their lives and to formulate identities). Moreover, these resources appear as various forms of discourses (including texts, voices and images). Individuals who can create, modify, interpret or even manipulate discourses can utilise such ideological resources (Mann 2012a). A few examples of these resources include the theory of political system, Confucianism and Christianity.

The institutions of ideological power sources can be classified into three subtypes, namely, transcendent, immanent and institutionalised. Transcendent ideological institutions, such as a Buddhist association in Taiwan, can transcend other existing ideological, economic, military and political institutions to create authority that can be deemed sacred (Mann 2012a). Alternatively, immanent ideological institutions intensify the existing cohesion, confidence and hence, the calibre of an established social group (Mann 2012a). An example of this subtype is the Chinese Communist procedure of cadre rectification, which was employed prior to 1949. Lastly, institutionalised ideological institutions are 'conservative and pragmatic, endorsing ideas, values, norms and rituals which serve to preserve the present social order' (Mann 2006: 348). This form is illustrated by dispatching party members to attend courses at the Party School of the Central Committee of the Chinese Communist Party.[4]

Second, the resources of military power sources can be found in '*the social organization of concentrated lethal violence*. "Concentrated" means mobilized and focused, "lethal" means deadly. Webster defines "violence" as "exertion of physical force so as to injure or abuse"' (Mann 2012a: xiii, original emphasis). An example of these resources can be a fleet of destroyers.

Military institutions are characterised by their concentrated and coercive forms and can provide an organisation reach that is intensive (in terms of the capacity to strictly organise and command a high degree of commitment from involved actors, regardless of a small or large coverage of areas or quantities), authoritative (in the sense of the capacity to give definite orders and to secure conscious compliance) and extensive (in the light of the capacity to organise large amounts of population over expansive territories to secure minimal stable submission and co-operation). In this manner, 'throughout most of history military striking range was greater than the range of either state control or economic-production relations' (Mann 2012a: 26). The hierarchy, structure and exertion of power within the organisation of an army can exemplify the abovementioned viewpoint.

Third, understanding the resources of economic power sources is relatively straightforward; that is, they are goods and services derived from mankind's efforts to extract, transform, distribute and consume the utility of the natural environment and humans (Mann 2012a). Basically, economic resources can satisfy the requirements for survival. However, the material improvement of living standards is largely based on the high increase and fair division of economic growth and resources. Furthermore, although the influence of technology on society may originate from ideological resources, it generally takes the form of military equipment (Mann 2006); economic productivity (Callinicos 2004) and administrative information storage; and tempo-spatial compression (Giddens 1987). Money, productive equipment and human labour are examples of such economic sources.

The institutions of economic power sources are the enduring patterned interactions of the processes of extraction, transformation, distribution and consumption of goods and services. Mann (2012a) calls these processes as circuits of praxis. Individuals who direct or even monopolise the circuits became a dominant class that can wield distributive and collective power as well as a 'distinctively stable, sociospatial blend of extensive and intensive power and of diffused and authoritative power' (Mann 2012a: 25). Such a dominant class can render the autonomy of the state relative. A few examples of these institutions of economic power sources are banking and monetary systems. Furthermore, the samurai comprised the dominant class under the domination of the Tokugawa Shogunate in Japan. Major samurai possessed lands and were the ruling class, whereas the ruled class was composed mainly of peasants (Moore 1966).

Lastly, states are the political sources of power according to Mann's (2012a) theory. The state is a political organisation that consists of a set of institutions and controls a certain area as its territory. Within this territory, the political organisation can exert effort to establish certain forms of political relationships with other societal forces. In other words, it can hold a certain degree of authority to formulate constraining rules, which are ultimately supported by organised armed forces.

The resources of the state sources of power are derived from 'the usefulness of territorial and centralized regulation' (Mann 2012a: 9). Regulation and, effectively, co-ordination are useful and become resources of the state because the territorial centralisation, organisation and governance of certain ideological, military and economic sources of power in society are necessary to achieve what could not be obtained otherwise. However, Mann (1988) warns that society's loss of control over territorial centralisation can lead to the increased despotic inclination of the state. In his theory, the state's despotic power pertains to 'the distributive power of state elites over civil society. It derives from the range of actions that state elites can undertake without routine negotiation with civil society groups' (Mann 2012b: 59).

According to Mann (2012a), a territorially centralised regulation produces two main characteristics in the institutions of state sources of power. First, the delimited territory of the state renders interaction with neighbouring states inevitable. Geography and geopolitics are two critical factors that influence the state. Thus, the grand strategies, military strategies and doctrines become essential for the state's security. Furthermore, geopolitics and domestic politics are intertwined in reality. For example, the Japanese invasion and brutalities during the Sino-Japanese War between 1937 and 1945 resulted in the rise of Chinese nationalism amongst peasants in Northern China (Johnson 1962).

Second, centralised regulation and governance indicates that the state extracts resources and absorbs and co-opts societal forces by establishing several institutions. An obvious example is the control of the state over its armed forces or a state-sponsored and -supervised religion. Such institutions can generate the capacity for the 'central state, despotic or not, to penetrate its territories and logistically implement decisions' (Mann 2012b: 59). This is the infrastructural power of the state in Mann's (2012b) account.

Notably, various 'conceptions of politics direct our attention to different levels and spheres of evidence' (Leftwich 2004a: 12). In other words, various definitions of politics lead to different objects, targets of research and data collection. In this monograph, the definition of politics used simply follows that of Leftwich (2004b), because it is sufficiently broad to satisfactorily incorporate Mann's (1988) definition of political/state power and Steinberg's (2001) theory of territorial political economy without distorting each conceptualisation.

Thus, defining politics in general and in a broad sense is now necessary to better grasp the characteristics of political-*qua*-state power in Mann's (2012a; 2012b) IEMP Model (IMES Model) and the relationship between politics and state power.

Leftwich's (2004b: 103) definition of politics is worth quoting,

> Politics comprises all the activities of co-operation, negotiation and conflict, within and between societies, whereby people go about organizing the use, production or distribution of human, natural and other resources in the course of the production and reproduction of their biological and social life. These activities are nowhere isolated from other features of life in society, private or public. They everywhere both influence and reflect the distribution of power, the structure of social organization and the institutions of culture and ideology in a society, or smaller groups within it. And all this may further influence and reflect the relations of a society (or a group or institution within one) with both its natural and

social environments, that is, with other societies or groups and institutions within them.

This definition views politics as a process. By definition, politics, thus, becomes an aspect of all human interactions. In other words, every human interaction involves politics. All human interactions are related to a certain level of co-operation, conflict and compromise in which people exercise power to use, produce and distribute resources. An obvious process of politics is that of two diplomats negotiating the terms of splitting an oil field that straddles the border between their countries or a couple deciding which film to see. Thus, defining politics in such a broad sense provides the political approach for interpreting and explaining human behaviour (Leftwich 2004b).

In Mann's (2012a) conceptualisation, political power is restricted to the state's power, whereas politics for Leftwich (2004b) can be observed in all human interactions. To absorb these two definitions, this monograph changes two terms from Mann (2012a), namely, political sources of power and political power, into state sources of power and state power, respectively. When the terms politics and political are used, they convey the meanings of Leftwich's (2004b) definition of politics.

Summary: The Argument Thus Far

This monograph introduced Steinberg's (2001) theory of territorial political economy, which points out that the use, regulation and representation of a certain geographical phenomenon can interact with one another and create the social constructions of the phenomenon. Such social constructions influence and are influenced by the physical conditions of the geographical phenomenon (nature) and the social practices, institutions and organisations that are conducted, established and formed, respectively, to transform and exploit the phenomenon (the material organisation of society).

Based on this theory, Steinberg (2001) builds the three primary models of the social constructions of the ocean: the Indian Ocean Model, the Micronesia Model and the Mediterranean Sea Model. The last one contains its original version and three variations, namely, Freitian, Grotian and Seldenian. The Indian Ocean Model constructs the ocean as a space of passages across which trade and migration can be undertaken. Moreover, it constructs the ocean as a void external to civilisations and state governance. In contrast, the Micronesia Model constructs the ocean as a claimable and possessable resource reservoir.

The Mediterranean Sea Model lies between these two extremes. Its original version constructs the ocean as an unclaimable space of arena on which concerned states can wield military forces to dominate marine areas. Similarly, the Freitian Variation concurs with the concept that states should govern sea lanes. However, it pays more attention to the interests of the state in governing the sailing routes of trade and commercial affairs. However, the Grotian Variation emphasises that the sea should be governed according to the natural law of the community of all states. Finally, the Seldenian Variation stresses the right of coastal states to claim coastal waters and the marine areas allocated for fisheries.

Moreover, this monograph employs the Steinbergist Classification together with the IMES Model, which constitutes the primary analytical framework of this monograph, guides the focus of observation and establishes the rationale for interpretation. As previously mentioned, the IMES Model purported by Mann (2012a; 2012b) urges students of power and social change to examine the four most important sources of social power: ideological, military, economic and state power.

Guided by the primary analytical framework composed of Steinberg's (2001) and Mann's (2012a; 2012b) concepts and theories, I will examine the social constructions of the ocean formulated by the ideological, military, economic and state sources of power on modern Taiwan. Through such an endeavour, I intend to answer the overarching research question, that is, how did the state and other power sources construct the ocean in the society of modern Taiwan between 1949 and 2016? By answering this question, this monograph can help scholars explore the politics of marine policy stipulation and implementation on Taiwan according to the perspective of Leftwich (2004b).

Several Important Situations of Society on Taiwan as the Context

After presenting the primary analytical framework for this monograph, this section introduces several important situations of society in the society of modern Taiwan. This introduction constitutes the context of this monograph. Corresponding to the IMES Model, I will cite certain political, ideological, economic and military situations in Taiwan. The discussion on the context could help enhance the understanding of the social constructions of the ocean in modern Taiwan's society. Now I begin with the political situation.

Politics as the Context

The central government of the ROC retreated to Taiwan in 1949 and governed Taiwan, Kinmen and Matsu in an authoritarian manner through the imposition of martial law, which was lifted in July 1987. However, the process of democratisation only began when the ballot of the first direct popular presidential election was cast in 1996, a half century after Taiwan Island was retroceded to China in 1945.

I do not intend to discuss the details of democratisation on Taiwan Island, as Dafydd Fell (2012) has offered a comprehensive and insightful analysis of this political phenomenon. Instead, this section endeavours to explicate that I have opted to view the first direct presidential election as the beginning of democratisation. In turn, this explication, which points out where the ultimate state power has been laid, provides a rough sketch of the political situation in modern Taiwan.

I follow the minimalist definition of democracy and the idea that the most important political power of the ROC has been conferred to the presidency. The year 1992 saw 'the start of the next era in the constitutional order' (Fell 2012: 46). The members of the Legislative Yuan were completely re-elected in December. A direct popular presidential election began to be advocated in the same year, which was then in fact held in 1996. Although the presidency appointed the highest position of the executive branch of the government, the power of the presidency was further increased in 1997 when the people realised that the

> stipulations that the President's appointment of the Premier no longer required legislative approval, expanding the number of legislators from 164 to 225, giving the President the power to dissolve the legislature (only when the Legislative Yuan passed a vote of no confidence on the Premier).
>
> (Fell 2012: 47)

Furthermore, the authority and power of the presidency were also rooted in the tradition of politics since 1948, because the promulgation of the Temporary Provisions Effective during the Period of Communist Rebellion 'concentrated power on the Office of the President' (Fell 2012: 44). This tradition became increasingly prominent after the central government of the ROC retreated to Taipei in 1949. The two leaders of the KMT, namely, JIANG Jieshi (also known as CHIANG Kai-shek) and his son JIANG Jingguo (also known as CHIANG Ching-kuo) became presidents during the era of the martial law. Thus, real political authority and power were vested in the presidency. Furthermore, the presidential dictatorship during the time of the martial law enhanced the role and power of the presidency in politics in modern Taiwan.

Therefore, this monograph suggests that the democratisation of the ROC can be dated back to 1996 when people in the society of modern Taiwan began to 'have the opportunity of accepting or refusing the men [and women] who are to rule them' (Schumpeter 1976: 268–270, cited in Fell 2012: 30). This monograph offers the core of the political context for understanding the social constructions of the ocean in modern Taiwan by clarifying why the year 1996 is considered the initiation of the democratisation of the ROC. This brief introduction is useful for readers in comprehending the discussion of this monograph, particularly Chapter 4.

Nonetheless, one may wonder how the KMT maintained its authoritarian rule for more than 50 years in the society of Taiwan. Did the KMT authoritarian rule survive over time due to pure coercion? What enabled the two dictatorial presidents to retain the support of the people in Taiwan for approximately 40 years before the termination of the martial law? A widely accepted answer indicates that the performance of the state in governing the market was key (Wade 2004).

Economy as the Context

This sub-section is beneficial especially to Chapter 5. To begin with, Wade's (2004) notion of governing the market refers to a state's promotion and cultivation of industrialisation and economic development within its territory. The KMT obtained and retained people's support for its regime before the lifting of martial law by facilitating a high rate of economic growth (8.9 per cent on average between 1951 and 1987; Fell 2012: 17). Moreover, the KMT government achieved an equitable rate of income distribution. 'For example, the Gini coefficient, which compares the income share of the richest 20 per cent against the poorest 20 per cent, fell from 0.558 in 1953 to 0.321 in 1970. The average annual income rose, in spite of continued population growth, from US$48 in 1952 to US$4,991 in 1987' (Fell 2012: 18).

The root of the success of the KMT in governing the market and promoting economic prosperity was land reform.

> From 1953 landlords were compelled to sell to the government any land in excess of 2.9 hectares, which the government then resold to tenant farmers. In return the landlords received government bonds and shares in government companies. The number of owner–farmer families in the farming population rose from 38 per cent to 60 per cent between 1949 and 1957.
>
> (Fell 2012: 17)

'By non-communist standards the reform was significant, redistributing 37 percent of total cultivated area or 320,000 hectares' (Ishikawa 1967: 312, cited in Wade 2004: 76).

The land reform of the KMT resulted in the growth of agricultural production. 'Agricultural production grew at 4.4 per cent a year between 1954 and 1967, faster than just about anywhere else in Asia' (Wade 2004: 76). In addition, the KMT regime promoted industrialisation. Across the 1950s, adequate foundations were created for the production of plastics, synthetic fibres, glass, fertiliser, cement, textiles and other industrial products.

> By the late 1950s a substantial industrial sector was in existence, with plants well beyond the cottage industry stage, backed by simple repair and maintenance enterprises and a small but fast-growing number of components suppliers. Manufacturing as a share of GDP reached 22 percent by 1960.
>
> (Wade 2004: 77)

Moreover, massive capital accumulation occurred in Taiwan throughout the 1960s. As Wade (2004: 88) indicates, 'the rate of gross domestic investment to GDP stood at about 20 percent, which only developing countries with large natural resource investment and a few others including China then equalled'. Additionally, 'manufacturing as a share of GDP rose from 22 percent to 37 percent, from a rank of twelfth among fifty-five middle-income countries to a rank of first equal with Argentina' from the early 1960s to the late 1970s.

Over time, the KMT regime intensified its endeavours to enhance the research and development capacity of new industrial sectors in Taiwan during the 1970s and 1980s. Two important examples are the establishment of the Industrial Technology Research Institute in 1973 and the Hsinchu Science-Based Industry Park in 1980. These two entities are located in close proximity to each other. Furthermore, the KMT government stipulated certain industry-specific policies in the 1970s and 1980s. The state dominated by the KMT promoted the steel, machine tools, automobiles, electronics and shipbuilding industries (Wade 2004).

A rosy picture of economic development continued until the mid-1990s. 'By 1996, it was estimated that real per capita GDP in the society of Taiwan had reached 56 per cent of that in the United States (U.S.) and about 66 per cent of that in Japan' (Booth 2011: 102). As pointed out by Yun-peng Chu and Gee San (2011), the information and computer industries largely contributed to the economic growth in the 1990s. In particular, this decade witnessed an increase in the production of equipment that is compatible with IBM personal computers. Corresponding to the increased demands for personal computer products worldwide in general and in the United States (US) in particular, manufacturers of Taiwan seized the opportunity to occupy the market and expand their production.

Nevertheless, the rosy picture before the mid-1990s became less promising in the latter half of the decade.

> In the decade from 1995 to 2005, GDP grew by around 54 per cent, so the average annual growth rate over the decade from 1995 to 2005 has been slightly above 4 per cent per annum. This is higher than the OECD average, but much slower than what was achieved in the decades between 1960 and 1990.
>
> (Booth 2011: 107)

The global slow-down in the economic activities in the information and computer technology industries affected Taiwan, which was a major exporter of computer products. Along with Singapore, Taiwan experienced negative growth in this industrial sector (Booth 2011).

However, a slower growth rate appeared in the post-2000 period. 'During the 1970s, 1980s and 1990s average growth was 13, 7 and 10 percent, respectively. From 2000 to 2007, the average rate was less than 5 percent and in 2001 alone it was an appalling negative 27 percent' (Chu and San 2011: 128).

At the same time, the comparative advantage of the Chinese mainland and Taiwan economies changed.

> Taiwan's shares of the product markets in the US, Europe and Japan shrank from 3.43 percent, 1.02 percent, and 4.3 percent in 1999 to 2.08 percent, 0.84 percent and 3.30 percent, respectively, in 2005. By contrast, however, the corresponding share for China increased from 7.89 percent, 2.60 percent and 13.93 percent in 1999 to 14.57 percent, 4.56 percent, and 21.21 percent in 2005.
>
> (Chu and San 2011: 129–131)

Nevertheless, as discussed by Chu and San (2011), the trend of descending economic growth in Taiwan began after the DPP took over state power through the victory of Chen Shuibian (also known as CHEN Shui-bian; hereafter, CSB) in the 2000 presidential election. CSB abruptly ceased the construction of the Fourth Nuclear Power Plant in the late 2000s and recklessly announced his provocative claim that one country lies on either side of the Taiwan Strait. CSB's political actions conducted at the starting years of his rule shook the confidence of business leaders in the predictability of the policies of the ROC. Moreover, these actions rendered the economic elites anxious about the cross-Strait relationship. The reason for this tension is that the political leaders of the DDP were likely to embark on an action that could lead to *de jure* Taiwan independence and war between the two sides of the Taiwan Strait (Chu and San 2011).

The political actions of CSB were harmful for Taiwan's economy but conducive to the fortification of Taiwanese national identity. Co-operating with several mass media companies, CSB's political actions not only signalled the enhancement of Taiwanese nationalism but also increased his popularity (Chu and San 2011). This situation is a characteristic of the ideological sources of power in the society of modern Taiwan.

Ideology as the Context

The subsequent lengthy introduction of the ideological context is mainly helpful for comprehending Chapters 3 and 4. The most prominent ideological issue in Taiwan is the Taiwan Independence Movement (TIM) propelled by Taiwanese nationalism. The TIM occurred after the first full re-election of the National Assembly (*Guomin Dahui*) in 1991 and the revision of Article 100 of the Criminal Code of the Republic of China in 1992, which fortified the freedom of speech (Fell 2012). This ideological issue was centred on the question of identity and related to '[w]hether Taiwan is a nation-state or whether Taiwan has any claim to be a nation-state and how Taiwan should relate to "China"' (Schubert and Damm 2011: 1). The following text describes the concepts of identity, ethnic group, nation and nationalism to elucidate the ideological context of this monograph.

The Barthian definition can be used to understand ethnic groups ontologically,[5] which denotes that an ethnic group is a group of people who culturally differentiate themselves from others. Moreover, no one-to-one relationship exists objectively between an ethnic group and a culture, because the people determine critical cultural proximities and dissimilarities between themselves and other out-groups (Barth 1969).

To a certain extent, ethno-cultural boundaries can be manipulated, such that people can employ cultural dissimilarities to create boundaries between themselves and others. Any ethnic boundary that is erected to differentiate us from them depends on the continual expression and validation of cultural differences. An ethnic group is less of a fixed cultural unit of people. Instead, it is produced and reproduced in socio-cultural interactions and perceived and accepted during self-identification (Barth 1969). In other words, ethnic identities constantly evolve and are concerned with self-differentiation from others.

Frederick Barth (1969) suggests that an ethnic group exists ontologically as a group of people who culturally differentiate themselves from others. Furthermore, Rogers Brubaker (2004) argues that ethnic groups and nations can be viewed as epistemological representations. In other words, ethnic groups can be defined as comprising people with 'widely shared ways of seeing, thinking, parsing social experiences, and interpreting the social world' (Brubaker 2004: 79).

An ethnic group emerges when people label their group and assign themselves as members of such a group. Effectively, the epistemological dimension of nations as well as ethnic groups offers various perspectives on the world. 'They are ways of recognizing, identifying, and classifying other people, of construing sameness and difference, and of "coding" and making sense of their actions' (Brubaker 2004: 81). The distinction between the ethnic identities of the aboriginal Taiwanese, the Hokio people, the Hakka people and *Waishengren* (Mainlanders or people from other provinces of China who came to Taiwan after 1949) in Taiwan can illustrate the abovementioned points.[6]

Ontologically, a nation can be defined as an ethnic group or groups mobilised by nationalism in a context that involves politics and the state. Furthermore, nationalism can be defined as an ideology that demands that an ethnic group or groups should establish and support their state. Lastly, nationalism transforms an ethnic group or groups into a nation.

Importantly, no objective one-to-one relationship exists between ethnic groups and nations, that is, an ethnic group can be split into two or more nations. This tendency occurs when members of a nation regard themselves as belonging to the same ethnic group and as being politically mobilised to support different political causes and states (or would-be states) at the same time. Similarly, a nation can be composed of different ethnic groups, given that these groups deem it necessary to establish, support or view the state as their own. When many Hakka persons and Mainlanders supported the TIM, scholars argued that they embraced the Taiwanese national identity and became Taiwanese as a nation at the same time.

In summary, ethnic groups and the nation are conceptual devices for recognising, identifying and classifying people from the epistemological perspective. However, they are also cultural-political groups from the ontological perspective. An ethnic group comprises a group of people who culturally differentiate themselves from others. A nation is an ethnic group or groups who are politically mobilised by nationalism.

Briefly providing the meaning of identity is imperative. Identity can be defined as 'an actor's experience of a category, tie, role, network, group or organization, coupled with a public representation of that experience; the public representation often takes the form of a shared story, a narrative' (Tilly 1996: 7). Thus, identity can be classified into various types, such as an ethnic identity, national identity and *alma mater* identity.

The KMT has followed a zigzag course since the early 1990s in dealing with the issue of ethnic and national identities in Taiwan. Evidently, 'the KMT has always maintained a stance that is open to ultimate unification with the mainland' (Wu 2011: 52).

> As a ruling party in post-election times [that is, after 1996], the KMT's China policy is 'realist', in the sense that it is determined by having to deal with the Chinese government [that is, the Chinese Communists] and to maintain a strong bargaining position vis-à-vis Beijing. Depending on the circumstances, the KMT's stance on unification can then be modified to be stronger or weaker. During election times, it is of paramount importance for the KMT to advocate 'centrist' positions to maximize its votes, and the party hence approaches the DPP/Green camp line on asserting Taiwan's sovereignty and the ROC's independence. As an opposition party, the KMT shifts to 'fundamentalism', that is, to Chinese nationalism and a strong pro-unification stance in order to secure the votes, of its core supporters.
>
> (Schubert and Damm 2011: 6)

Nevertheless, although the KMT maintained an ideological stance that is open to ultimate unification, the post-2008 period (at the time, the KMT regained presidential power) continued to witness the flourishment of the Taiwanese national identity. As Schubert and Damm (2011) point out, the notion that people on Taiwan Island are different from those on the Chinese mainland has become increasingly acceptable on the island since the second half of the 1990s.

'It may be that the issue of Taiwan's quest for both a cultural and national identity has already been settled in the hearts and minds of the Taiwanese people, including the political elite' (Schubert and Damm 2011: 1). This viewpoint may have exaggerated the effect and popularity of Taiwanese nationalism. However, the phenomenon it describes was certainly enhanced during and after the rule of CSB, who began his reign in 2000. 'Chen's "Taiwanization" agenda soon developed a high profile and determined the course of his second term as president, which started in mid-2004' (Schubert and Damm 2011: 6).

This monograph suggests that the policy of DPP and CSB regarding Taiwanese identity formation can be perceived as merely a part of the TIM. According to Ann Heylen (2011) and Stefan Fleischauer (2011), the TIM can be traced back to the post-war period between 1945 and 1949. The authors point out that the endeavour to formulate the Taiwanese national identity involved revisions of modern Taiwan history, which dates back to the occupation of the Dutch East India Company of this island. Nevertheless, several current historical studies in Taiwan use the history of post-war Taiwan as an important ingredient for formulating the Taiwanese national identity (Heylen 2011). However, these studies argue that distinct Taiwanese ethnic and national identities, which clearly differed from those of other Chinese people on the Chinese mainland, emerged during these post-war years (Fleischauer 2011).

According to Fell (2012), the KMT promoted national integration after the post-war years by disseminating the KMT version of Chinese nationalism. In this version, 'the KMT argued that the ROC was the sole legitimate government of all China and the protector of traditional Chinese culture. The ROC's sacred mission was to unify China by retaking the mainland' (Fell 2012: 22).

However, the endeavours of the KMT regime regarding the national integration of the people in Taiwan began to falter despite its control of the education, media, military and political systems and their potential utilisation to instil Chinese nationalism in the people of Taiwan. As Fell (2012) discusses, the internal and external political environments of Taiwan changed during the 1970s and 1980s. Thus, the efforts of the KMT towards national integration were hindered.

Externally, the ROC lost its seat in the United Nations in 1971, whereas the US established its official diplomatic relationship with the People's Republic of China (PRC) in 1979. The ROC continued to lose its diplomatic recognition after this loss. Moreover, the loss of formal allies undermined the credibility of the Chinese nationalism promoted by the KMT and its claim that the ROC is the only legitimate state of the whole Chinese nation. Moreover, as the PRC became the only recognisable state across China on the international stage, the KMT's 'slogan of retaking the mainland sounded more of an anachronism than ever before, even to many KMT politicians' (Fell 2012: 134–135).

Internally, in the 1980s, 'political liberalization and the rise of an opposition also meant that for the first time there was open debate over identity issues' (Fell 2012: 135). As the economy improved, and development was promoted, people from high educational and wealthier backgrounds became increasingly less receptive to the Nationalist discourses of the KMT. Thus, '[b]y the late 1980s, the Mandarin-only official language policy was being challenged by the DPP. Along with democratization came the demands of ethnic justice and self-determination' (Fell 2012: 135).

Therefore, as previously mentioned, the KMT adapted to this new environment and adopted an oscillatory tactic to preserve and promote its ideology. Specifically, it oscillated between realist and fundamentalist stances and aimed to, ultimately, achieve the re-unification of the Chinese mainland and Taiwan (Wu 2011).

Furthermore, as Fell (2012: 144) reports, the percentage of public opinion that supported Taiwanese independence, Taiwan's re-unification with the Chinese mainland and status quo reached 7.8 per cent, 39.5 per cent and 17.5 per cent, respectively, in 1992 and increased to 13.6 per cent, decreased to 22 per cent and increased to 45.8 per cent, respectively, in 1996. The trend continued to change. In 2002, the percentage of this public opinion on independence, unification and the status quo reached 18.1 per cent, 18.2 per cent and 51.2 per cent, respectively. Even when the KMT regained the presidency in 2008, the percentage of public opinion that supported independence (23.1 per cent) exceeded that of unification (10.2 per cent), whereas that for status quo was 57.3 per cent.

Moreover, as Fell (2012: 147) indicates, the percentage of public opinion on self-identification as Chinese was 12.5 per cent, 6.2 per cent and 4 per cent in 2000, 2004 and 2008, respectively, whereas as Taiwanese was 36.9 per cent, 41.1 per cent and 48.8 per cent in 2000, 2004 and 2008, respectively. The percentage as Chinese and Taiwanese was 44.1 per cent, 47.7 per cent and 43.1 per cent in 2000, 2004 and 2008, respectively. The percentage of self-identification as Taiwanese (45 per cent in 2005; 52.4 per cent in 2010) surpassed that of self-identification as Chinese and Taiwanese (43.4 per cent in 2005; 40.4 per cent in 2010) despite MYJ of the KMT holding the presidency. Notably, Fell's (2012) data correspond to the analysis of Schubert and Damm (2011) on the relationship between Taiwanese nationalism and the Taiwanisation policy of CSB between 2000 and 2008.

This sub-section introduces the most prominent ideological trend, that is, the TIM, which can be traced back to the post-war Taiwan society. Taiwanese ethnic and national identities, as described by Fleischauer (2011), emerged and developed since 1945. Although the KMT-dominated ROC exerted great efforts to accomplish the task of national integration of the people in Taiwan, the domestic and foreign environments generally altered the credibility of the Chinese nationalism promoted by the KMT after the 1970s. Since the initiation of democratisation in 1996, the pace and intensity of the formation of Taiwanese ethnic and national identities have increased. By 2010, an increasing number of people identified themselves as only Taiwanese instead of Taiwanese and Chinese (Fleischauer 2011).

This discussion introduced the ideological context for understanding the social constructions of the ocean in modern Taiwan. Notably, it does not address the relationship of the ROC with the Chinese mainland and the US. Nevertheless, the situation of the military sources of power involved the Chinese mainland and the US. The content of the next sub-section may help readers approach Chapter 4.

Military as the Context

The military sources of social power on modern Taiwan are a legacy of the Nationalist-Communist Civil War (1946–1949). Although the Chinese Communists won the civil war, they failed to prevent the Nationalist troops from retreating to Taiwan. The military skirmishes that ensured afterwards were initially related to the struggles between Chinese Communism and Chinese nationalism (the KMT version), whereas the military confrontation (especially that after the 1990s) was related to the attempt of the Chinese Communist to re-unify China.

As Kuo-cheng Chang (2013) indicates, CSB's support for the TIM strengthened Taiwanese ethnic and national identities. In other words, the opportunity for peaceful re-unification was diminishing, and the possibility of an all-out military conflict between the two sides of the strait was increasing. In fact, the military sources of power wielded by the ROC were dovetailed to cope with the military threat from the Communists after 1949.

To grasp the general condition of the military forces one year prior to the final year in which the analysis of this monograph ends, the following text introduces the *2015 National Defense Report, the Republic of China* (Ministry of National Defense 2015). Importantly, the report became the military policy of the ROC. This report demonstrates the role the overall ROC military forces played and the function of each branch of the forces. According to the second article of the National Defense Act, the military power of the ROC should aim 'at utilization of comprehensive national power to establish national defense military force, to assist disaster prevention and rescue, safeguard national security and maintain world peace' (Law and Regulations Database of the Republic of China, no date). This grandiose goal was intended to be implemented by the three branches of the military forces, namely, the army, navy and air force.

According to the report, the army was ordered to garrison the land and strategic islands in peacetime and to counterstrike any amphibious assault and military attempt intended to seize islands during wartime. Moreover, the navy was commanded to maintain the security of the sea areas of Taiwan during peacetime and to co-ordinate with other branches of the armed forces and eliminate enemy blockades and invasions during wartime. Lastly, the air force was tasked with the reconnaissance of the surrounding areas of Taiwan during peacetime and is responsible for dominating the air space and co-operating with other branches of the armed forces during wartime.

The functions that the military forces of the ROC intended to provide were a result of their perception of the geopolitical environment of the ROC. The analysis of the military forces on the geopolitical environment focused on the relations among Taiwan Island, the Chinese mainland, Japan and the US. Currently, the military forces of the ROC deemed that the military and diplomatic policies of the US are major factors that caused potential military conflicts in the Western Pacific Ocean. The reason underlying this notion is that the US intended to re-consolidate its military domination and alliances in East Asia. Moreover, the sovereignty disputes of Japan and countries in Southeast Asia with the PRC over the demarcation of marine area in the East and South China Seas introduced instability to the regional military security.

The military forces in the society of modern Taiwan followed the foreign policy of the state, which advocated the peaceful development of marine resources through co-operation with concerned states. Moreover, the military forces of the ROC observed the principle of maintaining regional peace and security and avoiding potential disturbances and conflicts. Nevertheless, the armed forces of the ROC adopted a strong sense of vigilance over the Chinese Communist military power during the ebb and flow of the cross-Strait relationship. For the ROC's military forces in the mid-2010s, the following capacity of the PRC's People's Liberation Army (PLA) should be considered.

First, the PLA acquired the ability to monitor the situations of the marine and air areas that lie west of the First Island Chain (FIC) in the West Pacific Ocean. Second, the PLA Ground Force mechanised and digitised itself. Thus, it obtained sufficient ability to seize the offshore islands of Taiwan. Third, the PLA Navy had fortified its abilities for military transportation and amphibious warfare as well as its nuclear and aircraft-carrying capabilities. The PLA Navy had obtained the capability to initiate nuclear counterstrikes, to deny foreign military interference and to control and blockade the Taiwan Strait. Fourth, the PLA Air Force had been able to undertake superior offensive and defensive air and outer space operations. The PLA Air Force had acquired several of the most advanced aircrafts, such as J-20 fighters, and H-6K bombers and missiles, such as *Hong Qi* (Red Flag) and the S-3000 series. In other words, it achieved air superiority in areas that lie west of the FIC.

Fifth, the PLA Second Artillery Force possesses nuclear and conventional missiles. Specifically, this military force could launch large-scale joint strikes and could deter foreign military power from interfering with the military operations in the Taiwan Strait and on Taiwan Island. Sixth, the PLA had proven that it can launch electronic warfare to paralyse the surveillance and the command-and-control systems of the armed forces of the ROC. Seventh, the PLA could infiltrate and corrupt the information systems of the military and government cyberspace of the ROC. In this regard, the PLA was capable of hacking and planting computer programmes to steal and transfer classified data. Eighth, the PLA had launched various military satellites into space and was able to conduct long-range precision strike missions in areas west of the FIC.

Based on the presented perception of possible military threats from the PLA, the military power of the ROC intended to enhance its defence through the following endeavours. First, the armed forces of the ROC aimed to cultivate and fortify the capacity of joint operations of the three military services.

Second, they attempted to construct and to enhance the capabilities of military innovation for conducting asymmetric warfare. 'These capabilities, when unleashed, shall be employed to attack enemy's critical weakness so as to delay, sabotage, or paralyze the enemy's operational tempo and capacities' (Ministry of National Defense 2015: 88). Third, they intended to defend crucial military infrastructure and weapon systems and aimed to reduce damages that might have been caused by the first sudden strike of the PLA and to improve the survivability of the military forces of the ROC.

Connections between the Ocean and the Aforementioned Context

As briefly discussed earlier, the ROC had exerted efforts to foster and govern an industrial capitalist system in the society of Taiwan since 1949. As a result, Taiwan's economy improved and increased dramatically. Meanwhile, the KMT's version of Chinese nationalism was promoted. However, the ethnic and national identities in modern Taiwan generally deviated from that originally contemplated by the KMT. Taiwanese national identity was fervently promoted after the democratisation of the ROC, which aggravated the cross-Taiwan Strait relationship, wherein Chinese Communists sought to re-unify China possibly through military means. The military forces of the ROC were well aware of this possibility and developed a set of counter strategies to defend the political regime in Taiwan. The subsequent discussion intends to enhance the understanding of the connections between the context and the social constructions of the ocean in modern Taiwan.

First, as Chang (2013) identifies, the military sources of power controlled by the ROC clearly understood that the navy as well as the air force were crucial for coping with the possible military aggression of the Chinese Communists. In addition, the navy of the ROC was expected to inflict much damage to the PLA and to largely increase the cost for the PLA to cross the Taiwan Strait. The navy was responsible for preventing amphibious operations against the ROC through the western coast of Taiwan Island and a surprise assault through the eastern coast.

Second, as Juntao Xie (1977) indicates, a sound maritime transport system needed to be developed to enhance and supplement the naval power. After the 1976 Six-year National Economic Construction Plan, the state developed the maritime shipping industries and commercial ports, whose related industries generated 371.4 million New Taiwan Dollars as of 2007. Taiwanese merchant carrier fleets ranked the fourth largest one in the world. Within the context of the state facilitating the economic growth in Taiwan, several maritime shipping corporations (that is, Evergreen Marine, Yang Ming Marine Transport and Wan Hai Line) became the top 20 largest maritime shipping companies worldwide. Meanwhile, the value of exports reached more than 60 per cent of the gross domestic product in 2006 (Chang and Chen 2009: 3). As such, maritime shipping and its related industries played a critical role in Taiwan's economy.

Third, in addition to the state's concentration on the development of the maritime shipping industries, intellectuals in Taiwan examined and provided policy advice for the state on maritime and ocean affairs. Many scholars focused on the ways and means in which people can utilise and derive from the ocean. Others paid attention to the formation of Taiwanese national identity by incorporating a few of the characteristics of the ocean into discourses on Taiwanese nationalism, especially since the late 1990s. Nevertheless, the contents of the discussions of intellectuals on the utilisation of the ocean, regardless of whether for economic gains or for identity formation, were related to politics at the national and international levels.

Fourth, as previously mentioned, the actions of the ROC in relation to maritime and ocean affairs could influence the relationships between the ocean and the military, which are the economic and ideological sources of power in the society of modern Taiwan. In general, the ROC (the state), after it withdrew to Taiwan, stipulated a series of legal acts and policies that aimed to foster the growth of the maritime and marine-related industries.

Against this background, this monograph provides an interpretation of the social constructions of the ocean formulated by the ideological, military, economic and state sources of social power in modern Taiwan primarily after 1949.

The Outline of This Monograph

To reiterate, this monograph intends to answer the following overarching research question: How did the state and other power sources of society construct the ocean in Taiwan since 1949? Moreover, the IMES Model and the Steinbergist Classification were employed as the primary analytical framework for collecting and interpreting data to answer the overarching question. Thus far, Chapter 1 clarified the question and elaborated on the framework.

Chapter 2 addresses the social construction of the ocean generated by the state in modern Taiwan. It outlines the trajectory of the transformation of the state's construction of the ocean, which was initially one following the Indian Ocean Model. It then absorbed the elements of the Freitian Variation of the Mediterranean Sea Model in terms of the maritime commercial activities in Taiwan after 1949. In the 1990s, the construction assimilated the elements of the Grotian Variation from international maritime treaties. Lastly, it incorporated the elements of the Seldenian Variation and the Micronesia Model.

Chapter 3 explores the social constructions of the ocean embedded in the written works of intellectuals in natural sciences, social studies and humanities in Taiwan chiefly after the 1980s. These works were published at a time when the publications on maritime and ocean affairs before this period are generally difficult to find and irrelevant for this monograph. Marine engineers who sought to prevent the erosion of the land by the sea construct the ocean similar to the Indian Ocean Model. Marine biologists, ecologists and engineers who intended to extract and accumulate knowledge about the ocean construct it similar to the Micronesia Model. Additionally, legal, political and economic specialists who advocated the attainment of legally just marine-territorial demarcations and sustainable development of marine resources construct the ocean in a similar approach. Furthermore, managerial scholars who aimed to maximise the profits of maritime shipping constructed the ocean according to the Mediterranean-Freitian fashion, whereas political scientists concentrated on sea power and constructed the ocean in a manner similar to the original Mediterranean Sea Model. Penultimately, economists who urged the state and fishing industry to follow international fishery regimes constructed the ocean according to the Mediterranean-Grotian approach. Finally, many scholars on humanity and literary writers tended to construct the ocean along the Indian Ocean Model in their written works.

Chapter 4 analyses the construction of the ocean according to the military sources of power in Taiwan. Three time periods form the structure of this chapter. The period between 1949 and 1971 witnessed that the ROC navy become more powerful than the PRC navy, whereas the KMT-dominated state did not surrender the grand strategy of retaking the Chinese mainland through military power. Moreover, the military's construction of the ocean during this era can be generally classified as the original Mediterranean Sea Model. During the period between 1971 and 1991, the ROC adopting a defensive stance. In other words, its military forces constructed the ocean similar to the Indian Ocean Model. After 1991, the military forces of the ROC constructed the ocean along the style of the original Mediterranean Sea Model, given that the grand strategy of the ROC was transformed into a coercively defensive one. As such, the grand strategies of the state and the altering equilibrium of martial prowess across the Taiwan Strait across periods could result in various constructions of the ocean by the military sources of power.

Chapter 5 interprets the social constructions of the ocean that originated from the economic sources of social power in modern Taiwan. The chapter is divided into two primary parts. Instead of evaluating the representations of the ocean offered by all marine-related economic power sources, the use of the ocean is examined through the most relevant maritime shipping and fishing industries using the lens of the Steinbergist Classification. In the first part, I examine the practices of several major maritime transport companies and suggest that their construction of the ocean mainly fell under the Indian Ocean Model. The second part investigates the practices and development of the fishing industries, whose original construction of the ocean could also be classified under the Indian Ocean Model instead of the Mediterranean-Seldenian Variation as Steinberg (2001) notes for European-American societies. However, the fishing industry in modern Taiwan gradually absorbed the elements of the Seldenian Variation after the late 1970s and assimilated the elements of the Grotian Variation after the mid-2000s.

Lastly, Chapter 6 presents the research findings: the social constructions of the ocean as formulated by the ideological, military, economic and state sources of social power in the society of modern Taiwan between 1949 and 2016 were mainly consistent with Steinberg's (2001) theory of territorial political economy. This monograph demonstrates the manner in which people or organisations intend or urge others to use and regulate the ocean, which can influence how the people or organisations construct the ocean. This notion corresponds to and complements one of the main theses of the theory, that is, the use, regulation and representation of a geographical phenomenon are interlinked to and influenced by one another, which creates a matching social construction of the phenomenon. Furthermore, this monograph illustrates the notion that the ocean, together with the material organisation of society, can influence the social construction of the ocean. Additionally, this monograph exemplifies the view that the social construction of the ocean can influence the material organisation of society and the first nature of the ocean. Nonetheless, understanding the social constructions of the ocean in Taiwan can be a steppingstone for the further exploration of the politics of the stipulation and implementation of the ROC marine policy.

Notes

1 I follow Michael Mann's (2012a; 2012b) IEMP Model to investigate the main topic of this monograph. However, because '[p]olitical power means *state* power' (Mann, 2012b: 9, original emphasis), I replace the uppercase 'P' of the IEMP Model with uppercase 'S'. Moreover, for ease of pronunciation, I alter the sequence of the capital letters 'E' and 'M'. Thus, I used the term 'the IMES Model' to refer to the IEMP Model.

2 Hereafter, these two laws are considered the two marine-territorial laws of the ROC or simply the two marine-territorial laws.

3 The IMES Model is complicated and contains at least three elements, namely, the four irreducible sources of social power as the main constituents of society; a theory of social change; and an eclectic approach to the structure–agency polemic. This monograph does not introduce the second and third elements, simply because it aims to explore the social constructions of the ocean instead of the broad phenomenon of social change in modern Taiwan. The IMES Model is used to identify the aspects that require analysis.

4 An implication of this definition is that culture can, thus, be defined as the ideational constructs of the people's moral values, norms and normative choices; perceptive concepts and theories; aesthetic judgements and productions and religious faiths and commitments.

5 In this monograph, ontology refers to the fundamental assumptions of scholars about the nature of the social and political world and the entity that exists within that world in reality (Hall 2003). Epistemology refers to theories about the knowledge and understanding of human beings about the world and existence.

6 In general, people believe that four main ethnic groups exist in the society of modern Taiwan, namely, the Hokio people, the Hakka people, *Waishengren*, and aboriginal Taiwanese people.

References

Abrams, Philip (1982) *Historical Sociology* (Somerset; Open Books Publishing).

Barth, Fredrik (1969) *Ethnic Boundary and Boundaries: The Social Organization of Culture Difference* (Oslo; Scandinavian University).

Booth, Anne (2011) 'Is the Taiwan Model of Growth, Human Resource Development and Equity Sustainable the Twenty-First Century', in Robert Ash, John W. Garver, and Penelope B. Prime (eds.), *Taiwan's Democracy: Economic and Political Challenges* (London; Routledge), 101–124.

Brubaker, Rogers, (2004) *Ethnicity with Groups* (Cambridge, MA; Harvard University Press).

Callinicos, Alex (2004) 'Marxism and Politics', in Adrian Leftwich (ed.) *What Is Politics?* (Cambridge; Polity Press), 53–66.

Chang, Kuo-cheng (2013) *Dong Ya Haichuan Lun* [An Essay on the Seapower in East Asia] (New Taipei City; Book Republic).

Chang, Chih-ching and Chen, Shiou-yu (2009) 'Taiwan Hangyun Maoyi yu Gouji Jingzheng' (Taiwan's Shipping, Trade and International Competition), in Wen-yan Chiau (ed.), *Haiyang Taiwan: Yongxü Fazhan: Haiyang Chanye yu Keji Chuangxin* (Oceanic Taiwan: Sustainable Development: Marine Industries Technology Innovation) (Taipei; Taiwan Research Foundation), 1–26.

Chu, Yun-peng and San, Gee (2011) 'Taiwan's Industrial Policy and the Economic Rise of the PRC: Opportunities and Challenges', in Robert Ash, John W. Garver, and Penelope B. Prime (eds), *Taiwan's Democracy: Economic and Political Challenges* (London; Routledge), 125–145.

Coast Guard Administration (2004) *Guojia Haiyang Zhengce Gangling* (National Oceans Policy Guidelines) (Taipei; Coast Guard Administration, Executive Yuan).

Council of Marine Affairs Advancement (2006) *Haiyang Zhengce Baipishu* (Ocean Policy White Paper) (Taipei; Research, Development, and Evaluation Commission, Executive Yuan).

Fell, Dafydd (2012) *Government and Politics in Taiwan* (London; Routledge).

Fleischauer, Stefan (2011) 'Perspectives on 228: The "28 February 1947 Uprising" in Contemporary Taiwan', in Gunter Schubert and Jens Damm (eds.), *Taiwanese Identity in the Twenty-first Century: Domestic, Regional and Global Perspectives* (London; Routledge), 35–50.

Giddens, Anthony (1987) *The Nation-State and Violence* (Berkeley; University of California Press).

Hall, Peter A. (2003) 'Aligning Ontology and Methodology in Comparative Research', in James Mahoney and Dietrich Rueschemeyer (eds.), Comparative Historical Analysis in the Social Sciences (Cambridge; Cambridge University Press).

Heylen, Ann (2011) 'Legacies of Memory and Belonging in Taiwan History', in Gunter Schubert and Jens Damm (eds.), *Taiwanese Identity in the Twenty-first Century: Domestic, Regional and Global Perspectives* (London; Routledge), 17–34.

Hou, Kuang-hao (2012) *State Domination in China: An Examination of the Applicability of the IMEP Model for the Analysis of Chinese Politics in the 20th Century* (Heidelberg; Heidelberg Press).

Ishikawa, Shigeru (1967) *Economic Development in Asian Perspective* (Tokyo; Hitotsubashi University).

Johnson, Chalmers (1962) *Peasant Nationalism and Communist Power: The Emergence of Revolutionary China: 1937–1945* (Stanford; Stanford University Press).

Law and Regulations Database of the Republic of China (no date) *National Defense Act* <law.moj.gov.tw/Eng/LawClass/LawAll.aspx?PCode=F0010030>.

Leftwich, Adrian (2004a) 'Thinking Politically: On the politics of Politics', in Adrian Leftwich (ed.), *What is Politics? The Activity and its Study* (Cambridge; Polity Press), 1–22.

Leftwich, Adrian (2004b) 'The Political Approach to Human Behaviour: People, Resources and Power', in Adrian Leftwich (ed.), *What is Politics? The Activity and its Study* (Cambridge; Polity Press), 100–118.

Leftwich, Adrian (2009) 'State-Business Relations – Concepts and Methods: Analysing the Politics of State Business Relations – A methodological Concept Note on the Historical Institutionalist Approach', Discussion Paper 23, Improving *Institutions for Pro-Poor Growth Publications*, retrieved from http://www.ippg.org.uk/publications.html, accessed on 19 July 2019.

Lukes, Steve (2005) *Power: A Radical View*, 2nd edition (New York; Palgrave Macmillan).
Mann, Michael (1988) *States, War and Capitalism: Studies in Political Sociology* (Oxford; Basil Blackwell).
Mann, Michael (2006) 'The Sources of Social Power Revisited: A Response to Criticism', in John A. Hall and Ralph Schroeder (eds), *An Anatomy of Power: The Social Theory of Michael Mann* (Cambridge; Cambridge University Press), 343–399.
Mann, Michael (2012a) *The Sources of Social Power, Volume 1: A History of Power from the Beginning to AD 1760*, 2nd edition (Cambridge; Cambridge University Press).
Mann, Michael (2012b) *The Sources of Social Power, Volume 2: The Rise of Classes and Nation-states, 1760–1914*, New edition (Cambridge; Cambridge University Press).
Ministry of National Defense (2015) *2015 National Defense Report, the Republic of China* (Taipei; Ministry of National Defense, Republic of China).
Ministry of Education (2007) *Haiyang Jiaoyu Zhengce Baipishu* (Ocean Education Policy) (Taipei; Ministry of Education, Republic of China).
Moore, Barrington (1966) *Social Origins of Dictatorship and Democracy: Lord and Peasant in the Making of the Modern World* (Boston; Beacon Press).
Neuman, W. Lawrence (2003) *Social Research Methods: Qualitative and Quantitative Approaches*, 5th edition (Boston; Allyn and Bacon).
North, Douglass (1990) *Institutions, Institutional Change, and Economic Performance* (Cambridge; Cambridge University Press).
Parsons, Talcott (1949) *The Structure of Social Actions: A Study in Social Theory with Special Reference to a Group of Recent European Writers* (New York; The Free Press).
Peters, B. Guy (1999) *Institutional Theory in Political Science: The 'New Institutionalism'* (London; Pinter).
Research, Development, and Evaluation Commission (2001) *Haiyang Baipishu* (Ocean White Paper) (Taipei; Research, Development, and Evaluation Commission, Executive Yuan).
Schubert, Gunter and Jens Damm (2011) 'Introduction', in Gunter Schubert and Jens Damm (eds.), *Taiwanese Identity in the Twenty-first Century: Domestic, Regional and Global Perspectives* (London; Routledge), 1–14.
Schumpeter, Joseph (1976) *Capitalism, Socialism and Democracy* (London; George Allen & Unwin).
Steinberg, Philip E. (2001) *The Social Construction of the Ocean* (Cambridge; Cambridge University Press).
Tilly, Charles (1996) 'Citizenship, Identity and Social History', in Charles Tilly (ed.), *Citizenship, Identity and Social History* (Cambridge; Cambridge University Press).
Wade, Robert (2004) *Governing the Market: Economic Theory and the Role of Government in East Asian Industrialization*, 2nd edition (Princeton; Princeton University Press).
Wallerstein, Immanuel (1974) *The Modern World-System: Capitalist Agriculture and the Origins of the European World-Economy in the Sixteenth Century* (New York; Academic Press).
Weber, Max (1978) *Economy and Society: An Outline of Interpretive Sociology*, edited by Guenther Roth and Claus Wittich (Berkeley; University of California Press).
Wu, Yu-shan (2011) 'The Evolution of the KMT's Stance on the One China Principle: National Identity in Flux', in Gunter Schubert and Jens Damm (eds.), *Taiwanese Identity in the Twenty-first Century: Domestic, Regional and Global Perspectives* (London; Routledge), 51–71.
Xie, Juntao (1977) *Haiyang Kaifa Lun* (On the Development of the Ocean) (Taipei; Youth).

2 The State's Construction of the Ocean in Modern Taiwan[1]

Introduction

This chapter answers the following question: How did the state construct the ocean in the society of modern Taiwan? As posited in Chapter 1, the state is a political organisation embodying centrality in the sense that political relations radiate to and from the political organisation to cover a territorially demarcated area over which the organisation establishes a certain level of authoritative and binding rules, which are supported by organised military forces (Mann 2012b).

Although states are often separated and elevated from other nonstate or private individuals or social groups, as Joel Migdal (2001) indicates, a state is a member of the society situated in its territory. The Republic of China (ROC), that is, the state governing Taiwan, Kinmen and Matsu, despite being considered an international pariah, is the most crucial political organisation in the society of modern Taiwan.

Knowledge about how the ROC constructed the ocean plays a critical role in linking together and providing an understanding of a number of important questions related to human-marine relationships and the development and implementation of marine policies in modern Taiwan.[2] Furthermore, the interpretive analysis presented later intends to buttress one of the main theses of Philip E. Steinberg's (2001) theory of territorial political economy. How an individual or an organisation intends, or urges others, to use and regulate the ocean can shape, how the particular actor represents and, then, constructs the ocean; this is the essence of the thesis that the use, regulation and representation of the ocean are mutually produced and shape how the ocean is constructed.

In this chapter, I interpret the marine and maritime laws and the two most prominent marine policy white papers published by the ROC between 1949 and 2016. Maritime law is the body of private law 'regulating the economic and commercial aspects of marine transport' (Steinberg 2001: 15). The maritime laws examined in this chapter include the Maritime Act and the Shipping Act. Marine law (or the law of the sea) refers to public law that governs the ocean, 'particularly with reference to the control of ocean-space

DOI: 10.4324/9781003291343-2

and access to its living and nonliving resources' (Steinberg 2001: 15). The marine laws discussed in this chapter include the Law on the Territorial Sea and the Contiguous Zone of the ROC, the Law on the Exclusive Economic Zone and the Continental Shelf of the Republic of China, and the Fishery Act. The first two marine laws are marine-territorial laws, as introduced in the first chapter. This dichotomy between marine and maritime law is also accepted by Taiwan's academic legal field (Lin 2007; Yin 2003).

The laws stipulated by a state can be viewed as the state's will regarding how the state would govern certain issues and problems within the jurisdiction claimed by the state and within the parameters of its power. Therefore, the ROC's marine and maritime laws are its proclamations regarding how it intends to govern issues and problems related to maritime and ocean affairs. The intentions and underlying attitudes surrounding the stipulation of these laws can be found in the minutes of law-making conferences of the Legislative Yuan, the legislative branch of the central government of the ROC. Interpreting these laws and the minutes through the lens of the Steinbergist Classification facilitates a novel understanding of the state's construction of the ocean.

The construction of the ocean by premodern Chinese states (before it was altered by the European imperialist forces in the mid-nineteenth century) can typically be classified as the Indian Ocean Model. The ROC between 1912 and 1949 continued to construct the ocean in this style. However, having been pulled into the international community, the Chinese state strove to align its legal system with those of Western states and absorbed elements of the Freitian Variation of the Mediterranean Sea Model regarding maritime economic sectors after it withdrew to Taiwan in 1949. Regarding its maritime activities, since the 1990s, the state has endeavoured to follow the international community, assimilating elements of the Grotian Variation.

After the mid-1990s, the state also incorporated elements both of the Micronesia Model from the international community regarding the territorialisation of the ocean and the exploitation of its resources and the Seldenian Variation concerning fishery. Nevertheless, the state's construction of the ocean in the past left an imprint on the current legal system. Consequently, the construction of the ocean formulated by the state in modern Taiwan contained various elements that are also found in the models and variations of the Steinbergist Classification.

Certain crucial motivations also buttressed the making of the ROC's Ocean policy and its corpus of marine and maritime law. These motivations included: first, the state's efforts to promote a capitalist economy on Taiwan; second, the state's intention to use and conserve marine resources; and third, under the rule of the Democratic Progressive Party (DPP), the state's desire to foster a national identity that seemingly supported the Taiwan Independence Movement (TIM).

The layout of this chapter is as follows. The next section introduces an interpretation of the maritime laws of the ROC. The third section presents an interpretation of the ROC's marine laws and ocean policy. Finally, in the fourth section, I argue that the overall historical development and the trajectory of change in the state's construction of the ocean demonstrate that the construction emerged as a hybrid version that mixed elements of the three main models of the Steinbergist Classification in varying proportions.

The Steinbergist Classification and the theory of the territorial political economy were introduced in the previous chapter. I will now briefly discuss the general construction of the ocean produced and inherited by the traditional Chinese dynastic states to exemplify Steinberg's (2001) concepts and theories. Overall, the traditional Chinese states' construction of the ocean can be classified as the Indian Ocean Model. However, this construction underwent a transformation between the Southern Song and early Ming periods (1172–1435) into the Freitian Variation. The emergence of this Freitian-style construction is consistent with the main argument of the theory of territorial political economy.

The loss of the Northern territory and the overland trading passages to the Jurchen people heralded monumental changes in the geographical environment and state-society structures for the Southern Song state. It became necessary for the state of the Southern Song to use the navy for defence along the Yangzi River and seacoast. The state also had to rely on maritime trade and employ naval forces to protect its international trade routes at sea. The state's construction of the ocean changed to the Freitian Variation, accordingly. When the state of the Ming dynasty regained Northern China and more sources of revenue on land, it also began to face increasing nomadic threats from the North. This initiated changes to the state's maritime and naval policies and practices that led to the reconstruction of the ocean according to the Indian Ocean Model (Cole 2010; Lo 2012; Zheng 2012).

This example demonstrates that a state's construction of the ocean responds to the particular geographic-natural environment and the material organisation of society (that is, the social practices, institutions and organisations, as well as the technologies, employed to transform and exploit the natural environment), in which the state is situated, as Steinberg's (2001) theory of territorial political economy elucidates.

The traditional construction of the ocean by the premodern Chinese states, including the exception that prevailed between the Southern Song and early Ming eras, exemplifies the theory of territorial political economy.[3] I now follow the rationales of the theory of territorial political economy and the Steinbergist Classification to analyse the relationships between modern Taiwan's social structures and geographical environment on the one hand and the laws and policy of the state on the other.

The Maritime Laws of the ROC

The ROC inherited attributes of the Indian Ocean Model from the aforementioned traditional constructions of the ocean. In this section, I show that, regarding maritime activities, the state changed the inherited Indian-Ocean-style construction to absorb elements from the Freitian Variation of the Mediterranean Sea Model between the early 1950s and the 1980s. This shift was primarily a result of the state's attempt to foster a sound seaborne transportation system and to create an industrialised capitalist economy in Taiwan. Since the 1990s, the construction has changed again, assimilating elements from the Grotian Variation of the Mediterranean Sea Model, which has facilitated Taiwan's integration into the international economic system.

The trajectory of the changing ROC construction of the ocean began from the Indian Ocean Model. As Cole (2010) indicates, the state of the late Qing dynasty (1839–1912) attempted to strengthen its naval power for the purpose of defending China against imperialist encroachment. The Self-strengthening Movement organised by the Qing state after 1864 established four modern naval fleets. However, according to Cole (2010), the state was unable to coordinate inter-fleet operations. Nor did the state establish an effective central administration, logistical support or command-and-control institutions for its navy. Neither did the Qing state have a military-naval strategic goal or a vision for modern naval development.

As Cole (2010) points out, the Qing state acquired modern naval weapons and equipment, but did not establish the appropriate institutions, military doctrines or software for its navy. The Qing navy lost a decisive sea battle against imperialist Japan. The Qing state had merely purchased Western naval weapons to intercept threats along its coastal waters. For the Qing state, the ocean was considered a great void across which the imperialist states projected their military power onto China. This construction is shown by the apropos phrase summarising the Self-strengthening Movement, 'Chinese Learning as the Foundation, and Western Learning for Practical Use'. The poor performance of the Qing navy demonstrated that the state retained its old institutions of maritime and naval activities and the traditional construction of the ocean: that of the Indian Ocean Model.

The navy in the early Republican period (1912–1928) consisted almost entirely of the remains of the Qing state (Cole 2010). Even after the Chinese Nationalists (Kuomintang or KMT) completed their Northern Expedition and nominally dominated China in 1928, no significant improvement in naval power took place. The Japanese navy's amphibious assault on Shanghai in 1932 indicated that China remained unable to defend itself against this type of foreign military aggression. That the KMT did not (and could not) make efforts to rebuild or strengthen its navy and that they single-mindedly continued land-based military campaigns on the Chinese mainland demonstrate that the state's construction of the ocean largely remained that of the Indian Ocean Model.

Before losing the civil war against the Chinese Communists and withdrawing to Taiwan in 1949, the Nationalist-controlled state had not given any significant attention to the ocean, concentrating instead on the mainland. Although the state's practices regarding the ocean retained the old characteristics, how this was manifested, for example, the statutory laws, gradually showed signs that its construction of the ocean was changing. The Maritime Act was first promulgated in December 1929 when the KMT claimed their nominal unification of China and was then completely revamped in July 1962, although the actual process of the complete amendment began in March 1958 (Legislative Yuan 1958).

Holding Taiwan as the ultimate bastion against the Chinese Communists, the ROC faced immense threats from across the Taiwan Strait, and the state relied on the protection of the United States (US) Navy after July 1950 (Chang 2001: 527–530). In addition, the Nationalists, especially JIANG Jieshi (also known as CHIANG Kai-shek), understood that their loss to the Communists on the Chinese mainland was indirectly attributable to the failure of land reform before the post-Second World War economic recovery. Thus, the state was committed to promoting economic growth and to increasing its revenue in Taiwan (Wade 2004).

After 1949, the ROC embarked on the mission of establishing capitalism in Taiwan, as Wade (2004) indicates. The cultivation of capitalism by the state of a developing country entails industrialisation and economic development through industrial policies as well as utilising development-oriented economic bureaucrats and experts to govern the markets. Moreover, states pursuing capitalist development ultimately aim to catch up with the economic strength of the developed Western countries and engage in economic and trade relations with the wealthier economies with autonomy and equality.

The ROC began to implement industrial capitalism and facilitate economic development through a series of Four-Year Economic Plans. According to Wade (2004), during the First Four-Year Plan (1953–1956), the state allocated most of its available resources to agriculture, fertilisers and textiles. Starting with the Second Four-Year Plan (1958–1961), the economic strategies became increasingly sophisticated. Targets for the overall growth of national income and investment were established and heavy industries were promoted.

According to Wade (2004), the industrial development policy focused on export products that had high income elasticity and low transport costs. Policies of import substitution and export-oriented industrialisation were also implemented in the standard sequence: import substitution preceded export expansion. The substitution contributed substantially to Taiwan's economic growth from 1955 to 1960. In addition, the ROC initialised its support for trade, strengthened incentives for exports and promoted supplies and more direct foreign investment in 1958. State-led industrialisation and economic development persisted from the early 1950s well into the 1980s (Wade 2004).

The state also began to amend its Maritime Act in 1958[4], during its period of substantive promotion of economic growth. As the then legislator HE Zuozhi declared in 1958,

> in order to build a rich and strong country, the shipping industries must be developed. Therefore, we need to amend the Maritime Act…in order

> to make the industrial and commercial sectors prosper, we cannot rely on only the limited domestic market and have to increase exports and acquire foreign exchange. Therefore, we have to amend this act.
>
> (Legislative Yuan 1962: 127)

The connection between the state's promotion of economic growth and the amendment of the Maritime Act in 1958 lay in the state's belief that up-to-date maritime rules would enable adequate state regulation of the transport of goods, which would consequently facilitate a well-functioning maritime transport system. In addition, the system would contribute to lower transport costs and to Taiwan's overall economic development.

The exertion of state power to regulate international trade via sea lanes and establishing a sound maritime transport system could also be found in the Shipping Act of 1981 and its newer versions. Article 1 of the 1981 Shipping Act clearly stated that the purpose of this act was to strengthen the shipping system and develop the shipping industries and the economy (Legislative Yuan no date 1).

Furthermore, this act required the Ministry of Transportation (MOT) to coordinate and facilitate the development of the shipping industries. Article 14 requested that the MOT devise long-term plans for developing the shipping industries according to the trend of import-export trade and evolving international shipping standards. Similar demands can also be found in Articles 23 and 25. Moreover, in Article 26, the MOT was required to protect the interests of the shipping industries and to coordinate with the Ministry of Economic Affairs to consider undertaking reciprocal actions when its interests were harmed by foreign governments and shipping companies (Legislative Yuan no date 1).

In addition to the task of establishing a sound seaborne transport system through the Maritime Act and the Shipping Act, the state also attempted to construct a relatively coherent, stable and equitable legal framework for maritime activities. Ideally, constructing such a framework for economic activities in a capitalist society would serve the long-term interests of the overall economic system, preventing self-defeating, collective, irrational actions of individual capitalists pursuing immediate, self-serving interests (Weale 2004).

Article 19 of the 1981 Shipping Act, for example, authorised the MOT to regulate the number of ships to avoid the oversupply of freight and excessive passenger transportation capacity. Articles 27, 29 and 30 required all shipping companies to make their rates for freight and passenger transportation transparent and to demonstrate valid licences issued by the state in their advertisements. Discriminatory rates and illegitimate surcharges were forbidden.

Article 36 of the 1981 Shipping Act clarified the legal liability of smuggling. As the then legislators AI Shi and CHU Jiachang indicated, before the 1981 Shipping Act was stipulated, shipping companies that did not commit illegal activities had to bear the liability for their seafarers' private smuggling activities (Legislative Yuan 1981). This act thus removed the liability and costs previously incurred by the shipping companies for the illegal smuggling of their crew members, reducing the disincentives against undertaking legal shipping activities.

The 1962 Maritime Act had the same motivations: to promote economic growth, establish a sound seaborne transport system and construct a coherent and equitable legal framework. For example, the section in the act on maritime liens clarified the question of entitlement to a preferential right of compensation and castigated the disadvantages against the creditors of ships (Legislative Yuan 1962; Lin 2007). This act, thus, reduced the risk of financial lending to the shipping industries, facilitated the flow of credit in the industry and promoted the development of shipping businesses (Legislative Yuan 1962). Many other measures with similar effects were also updated and stipulated in the 1962 Maritime Act (Legislative Yuan no date 2).

Notably, in this chapter, I do not provide a detailed legal discussion of the Maritime Act and Shipping Act as this is beyond the scope of this chapter. Nor do I suggest that passing legislation necessarily leads to its proper implementation or that it will necessarily achieve its objectives. However, this chapter does suggest that, as demonstrated by the two aforementioned acts, the ROC had gradually changed its construction of the ocean, absorbing elements of the Freitian Variation before the 1990s.

The state endeavoured to promote economic growth by building a sound maritime transport system, and by establishing a coherent, stable and equitable maritime legal framework to facilitate that system. The ocean *per se* was not the target of regulation; rather, the state was to govern the use of the ocean surface for international maritime transport and trade through the legal framework. This was a key attribute of the Freitian Variation.

The intention of building a coherent and equitable legal framework for Taiwan's maritime shipping industries can also be found in the 1999 Maritime Act. The section on ships' crew was separated from the Maritime Act and became a more coherent statutory law in its own right in June 1999. Article 4 of the Maritime Act regarding the exercise of precautionary proceedings was amended to strengthen the protection of the maritime transport companies and related financial and loan sectors (Lin 2007; Legislative Yuan no date 2). Moreover, other amendments of the act, including those regarding the ownership of ships, maritime liens, ship mortgages, carriages, collisions of ships, salvage and general average also aimed at establishing a more coherent and equitable legal framework for the shipping industries in the society of Taiwan (Legislative Yuan no date 2).

Nevertheless, the amendments to the Maritime and Shipping Acts in the 1990s demonstrate that the ROC then dovetailed its maritime laws according to relevant international treaties and accepted the elements of the Grotian Variation regarding seaborne trade and transportation. The general Western construction of the ocean in the industrial capitalist epoch was 'characterised by a complex ocean-space construction that blended several tendencies'

(Steinberg 2001: 112). Regarding international maritime trade and transport, the ROC's construction of the ocean maintained the Mediterranean-Grotian elements.

As Steinberg (2001) identifies, this type of construction pervaded the community of states once the industrial capitalist era commenced in the nineteenth century, becoming an international norm. The deep sea was viewed as a great void or space across which goods and passengers could be transported, and maritime activities were subject to the rules of the international community only in a weak Grotian pattern. It was in the interest of industrial capitalism worldwide that the community of states considered the deep sea as a free space subject to weak international management and little security that would have reduced the risk of maritime activities (Steinberg 2001).

The ROC, as Wade (2004) indicates, still pushed for economic development through industrial policies after the 1990s. Biotechnology and finance were two areas that the state endeavoured to develop in the 1990s and early 2000s. Although free trade and economic liberalisation had gradually gained support from state organisations governing and managing the economy, entry into the World Trade Organisation (WTO) in November 2001 did not prevent it from 'acting to nurture some industries to competitiveness' (Wade 2004: xl). Furthermore, although Taiwan's representatives at the WTO surrendered certain state institutions that had coordinated and fostered economic development, the ROC still aimed to transform Taiwan into the Asia-Pacific Operation Center (APOC). This transformation required an updated legal framework governing Taiwan's maritime and shipping activities.

Therefore, with the intention of governing maritime activities to promote further economic growth, the ROC's construction of the ocean in the 1990s and early 2000s still fell under the purview of the Mediterranean Sea Model. Taiwan's efforts to absorb the new rules of international treaties and institutions to update its maritime legal framework for establishing the APOC resulted in a construction of the ocean that not only embraced the Freitian Variation but also incorporated the essence of the Grotian Variation regarding its maritime activities.

The international treaties that were absorbed and adopted by the ROC included the 1967 International Convention for the Unification of Certain Rules Relating to Maritime Liens and Mortgages (influencing Articles 26 and 29 of the 1999 Maritime Act of the ROC); the 1968 Hague Visby Rules (influencing Articles 54, 56, 60, 64, 69, 70, 71, 75, 76 and 131); the 1974 York Antwerp Rules (influencing Articles 110, 117, 119 and 120); the 1976 Convention on Limitation of Liability for Maritime Claims (influencing Article 21) and the 1989 International Convention on Salvage (influencing Article 103) (Legislative Yuan no date 2). The amendments to the Shipping Act of 1991, 1995 and 2012 showed the same influence of international treaties and norms. According to the Database Retrieval System of the Legislative Yuan (no date 1), all of these amendments aimed to align the Shipping Act with WTO requirements, for example, Articles 42 and 47.

To summarise the argument thus far, I have shown in this section that regarding maritime activities, the ROC's construction of the ocean absorbed elements of the Freitian Variation between the early 1950s and late 1980s. Since the 1990s, the construction has gradually changed and assimilated elements from the Grotian Variation. As demonstrated by the statutory laws of the Maritime and Shipping Acts as well as by the intention to stipulate and amend these acts, the state's changing construction of the ocean was affected by the geographical and geopolitical environment as well as by the state-society structure and relations.

After withdrawing to Taiwan, the ROC promoted economic development and industrial modernisation. Maritime trade and transport were considered essential to this policy objective. Therefore, the state deemed it necessary to establish a sound seaborne transport system and to regulate and manage the maritime activities of Taiwan. When the state pushed for further economic growth in the 1990s and recognised the changing international economic structure of which Taiwan was a part, the ROC attempted to build the APOC and altered its construction of the ocean by updating its relevant maritime laws according to the relevant international treaties. This echoes Steinberg's (2001) theory of territorial political economy.

The Marine Laws and Ocean Policy of the ROC

The law of the sea, or marine law, stipulated by the state on Taiwan after 1949 shows a similar trajectory of the changing construction of the ocean. However, in the state's governing of Taiwan's fishing industry, the construction shifted from the Indian Ocean Model towards the Seldenian Variation of the Mediterranean Sea Model. For economic needs, the state began accepting the legal rules of the international community in the 1990s. The promulgations of the 1991 Fishery Act and the two 1998 marine-territorial laws demonstrate that the state has absorbed elements of the Mediterranean-Seldenian construction and the Micronesian-type construction from the international community. However, the assimilation of the Micronesia Model also served the ideological function of national identity formation promoted by the state under the then president, CHEN Shuibian (also known as CHEN Shui-bian; hereafter CSB) of the DPP, which strove to establish a new, independent state of Taiwan recognised by the international community.

The Fishery Act was stipulated in December 1929 (Legislative Yuan no date 3). The original Fishery Act and the original Maritime Act, stipulated in the same year, represented the beginning of the change in the state's construction of the ocean in the late 1920s and early 1930s, even though the state on the Chinese mainland still largely retained its construction of the Indian Ocean Model. Similar to the Maritime Act, the Fishery Act underwent a complete overhaul when the state attempted to promote the economic development of Taiwan.

Nevertheless, it can be argued that the state's construction of the ocean over the issue of fisheries was not consistent with the Seldenian Variation. The 1970 Fishery Act aimed to regulate and manage the fishing industries on modern Taiwan rather than the fishing areas of the sea itself. The 1970 Fishery Act did not demonstrate the state's intention to take control of any portion of the ocean's fishing grounds. The process of amending the original Fishery Act began in October 1968 (Legislative Yuan 1968). The construction of the ocean evidenced by the modified act could still be interpreted as aligned with the Indian Ocean Model.

As the reputable economic technocrat LI Guoding (the minister of economic affairs at the time) clarified, the development of Taiwan's fishing industry, improvement of the people's diet and to establish an internationally competitive fishing industry, it was imperative to establish a relevant, adequate and updated legal framework (Legislative Yuan 1969a). In his report to the Legislative Yuan, LI Guoding also explicated three critical objectives that the amendment aimed to achieve: first, improve the state's governance of the fishing sector; second, promote the industrialisation of the fishing sector and third, protect poor fishermen (Legislative Yuan 1969a).

Additionally, although Article 10 of the 1970 Fishery Act stated that fishing rights were viewed as a right *in rem* (Legislative Yuan no date 3), the state demonstrated no sign of deeming fishing grounds as claimable territories or as areas comparable to the Exclusive Economic Zones (EEZs), which will be discussed later. The main focus was on fishing activities, and the primary goal was the development of the fishing industries as well as the overall modernisation of Taiwan.

The 1970 Fishery Act demonstrates that the state still constructed the ocean according to the Indian Ocean Model, as the state's interest in regulating fishing activities was linked to the overall goal of the economic development of Taiwan. One can find no clues that the state in the 1970 Fishery Act claimed fishing grounds as its exclusive possessions vis-à-vis other states. The 1970 Act was silent about the ocean *per se*, the control of the ocean and access to its other living and nonliving resources. This point is also evident in the then legislator, WEI Shouyong's speech: 'The foundation [of the Fishery Act] concentrates on management and registration … The main purpose of this act focuses on the development of the fishing sector' (Legislative Yuan 1969b: 7).

The Fishery Act was later completely upgraded; it absorbed elements of the constructions of the ocean embedded in the international treaties of the late 1980s and early 1990s. The 1991 Fishery Act was drafted in 1989, although the process had begun in 1987. According to YU Yuxian, the then Chairman of the Committee of Agriculture responsible for agricultural and fishing affairs, the amendment of the Act was critical, primarily because most of the coastal states worldwide had already proclaimed their EEZs of 200 nautical miles, in accordance with the United Nations Convention on the Law of the Sea (UNCLOS). The policies of neighbouring countries made it difficult for the fishing sector of Taiwan to expand and thrive without clashing with their fishing interests (Legislative Yuan 1991). Therefore, the ROC had to act in response.

The 1991 Fishery Act was not the only reaction by the ROC. The state's enactment of the two marine-territorial laws effectively incorporated the tenets of the international law of the sea into its legal frameworks. Therefore, the principles of the general construction of the ocean embedded in international law were integrated.

The construction of the ocean formed by UNCLOS showed a 'codification and institutionalisation of a multifaceted ocean-space construction' (Steinberg 2001: 149). Regarding the high seas, UNCLOS maintained the construction belonging to the Grotian Variation of the Mediterranean Sea Model. Concerning the EEZs and contiguous zones, UNCLOS materialised the construction of the Seldenian Variation of the Mediterranean Sea Model. Regarding the territorial seas, UNCLOS institutionalised the construction of the Micronesia Model (Steinberg 2001).

The UNCLOS became widely accepted. Japan and South Korea proclaimed the establishment of their EEZs in 1996. Negotiations regarding the questions of fishing areas and territorial sea delineations were undertaken among the Chinese Communist regime, Japan and South Korea in the same year (Legislative Yuan 1996a). To protect the interests of Taiwan's fishing sector, which had suffered from occasional harassment and intimidation by the police and military forces of neighbouring states after the 1980s, the ROC began to compose the two marine-territorial laws in 1989 (Ministry of the Interior 2002).

As the then vice-minister of foreign affairs, CHEN Xifan, identified, the necessity of stipulating these two laws lay in the state's ineligibility to be a party to UNCLOS. The executive branch of the ROC thus required legal authorisation that was in line with UNCLOS to undertake actions to protect fishermen and other legitimate resources within the ROC's EEZ (Legislative Yuan 1996a). Therefore, the Micronesian and the Mediterranean-Seldenian elements embedded in UNCLOS were absorbed by the state following the enactment of the two marine-territorial laws. For example, Article 2 of the law regarding the ROC's territorial sea was from UNCLOS' Article 2; Article 3 was from UNCLOS' Article 3; Article 4 from UNCLOS' Articles 6, 7 and 14 and Article 6 was from UNCLOS' Article 15. The law regarding the ROC's EEZ and the continental shelf was adopted from UNCLOS' Part V and VI (Fu 2001).

The composition of the first official ocean policy white paper began in 1998 and was published in 2001. The second edition of the white paper was published in 2006. The first edition contains six themes. The first theme of the first edition concerns marine-territorial security, which mainly regards the development of a naval force, the protection of rights for using the ocean and its resources and the eradication of criminal activities at sea. The second theme covers the sustainable development of oceanic living, non-living, marine tourism and shipping and transport resources and facilities. The third theme entails oceanic environmental protection and resource conservation. The fourth theme explicates the policy of coastal area management, and the last theme addresses the need to promote ocean studies and national ocean-oriented culture. Notably, the second edition of the white paper elaborates more on these six themes compared with the first.

The 1991 Fishery Act, two 1998 marine-territorial laws and two ocean policy white papers all demonstrate that the ROC, by the 1990s, had incorporated elements of the Seldenian Variation and the Micronesia Model, especially regarding the possession of the territorial sea and the use of living and non-living marine resources. That is to say, in the 1990s, facing changing domestic and international structures yet continuing to strive for the economic prosperity of Taiwan and to maximise the utility of exploiting its geographical environment, the ROC altered its construction of the ocean and then assimilated elements of the new construction of the ocean from the international community.

Thus, the trajectory of the state's construction of the ocean changed from the Indian Ocean Model before 1949 to moving closer to the Freitian Variation of the Mediterranean Sea Model before the late 1980s. In the 1990s, the state's construction of the ocean largely incorporated elements of the Grotian-style construction regarding maritime activities, of the Seldenian-style construction regarding fishery and of the Micronesian-type construction regarding the possession of the territorial sea, from the international community and treaties. The causes of the change in the state's construction of the ocean correspond to Steinberg's (2001) theory of territorial political economy. How an organisation or a person (in this case, the ROC) intends to use and regulate the ocean influences how that actor represents the ocean and, in turn, shapes its social construction of the ocean.

Nevertheless, the pursuit of economic wellbeing was not the only reason that the state absorbed elements of the Micronesia Model embedded in UNCLOS. This shift also served the ideological purpose of national identity formation on Taiwan. The process of enacting the two marine-territorial laws was marked by considerable controversy, as evidenced by the high number of drafts—ten—of the legislation (Legislative Yuan 1997). The controversy focused on the demarcation of the ROC's territory and the scope of its territorial sea. The stipulation of the marine-territorial laws in accordance with UNCLOS involved questions regarding the state's sovereignty, the TIM and Taiwanese national identity.

The then DPP legislator, LIN Zhuoshui, presenting his own draft for the Law on the Territorial Sea and the Contiguous Zone of the ROC, demanded that the scope of the territorial sea be limited to Taiwan and its pertinent islands (Legislative Yuan 1996a; 1996b). Lin and the DPP attempted to use the two marine-territorial laws to strengthen the TIM. However, his draft was eventually dropped; a similar attempt can be found with the two ocean policy white papers.

The ROC, governed by CSB (his governing lasted from May 2000 to April 2008), in its shift to absorbing elements of how the ocean was constructed that were embedded in UNCLOS, also began to construct the sea as a means of building Taiwanese national identity by proclaiming the need to cultivate a so-called ocean-oriented culture and to develop the oceanic and maritime characteristics of Taiwan. Maritime activities that were regulated and promoted by the state had been identified as necessary for economic development since the 1950s. However, neither white paper had ever clearly defined an ocean-oriented culture or what Taiwan's oceanic and maritime characteristics were.

Regardless of the level of the actual efforts made by the state in the 2000s to formulate a Taiwanese national identity based on so-called oceanic and maritime characteristics, the ROC gradually recognised the importance of the ocean. The ocean played a pivotal role in formulating a new national identity promoted by the DPP-dominated state.

The ocean remained essential for further economic development, especially after the state had assimilated elements of the Micronesia Model and acknowledged the benefits and resources that the population could extract from the ocean. This role of maritime resources for economic development became accentuated after May 2008 when the KMT-nominated MA Yingjiu (also known as MA Ying-jeou; hereafter MYJ) electorally regained presidential power.

The slogan, 'Blue Revolution: Developing the Country with the Ocean', used in MYJ's ocean policy manifesto during the 2008 presidential election campaign, highlighted this economic function of the ocean. The state, under an anti-TIM president, sought to incorporate elements of the Micronesia Model for economic progress, diminishing the potential effect of the ocean on promoting a pro-TIM national identity. Under the KMT, the state's construction of the ocean reflected its economic function rather than its ideological, discursive role.

Above all, it can be noted here that, as indicated by the theory of territorial political economy, changes in the material organisation of society (including state-society relations) and in the geographic-natural environment can result in alterations in the state's construction of the ocean (Steinberg 2001). The enhancement or decline of Taiwanese national identity certainly led to changes in state-society relations in Taiwan. However, influenced by these changes, it remained unknown whether the incorporation of the elements of the Micronesia Model could generate a more distinct ocean-oriented culture. The next chapter suggests that the incorporation failed to do so before 2016.

Conclusion

In this chapter, I have delineated the trajectory of the ROC's changing construction of the ocean. Steinberg's (2001) theory of territorial political economy indicates that the construction of a geographical phenomenon can be changed when either the material organisation of society that is institutionalised to exploit the particular phenomenon, or the phenomenon itself, changes. The altered construction can, consequently, modify the interactions between society and the geographical element. The modified interactions, in turn, lead to transformations in the material organisation of society and the geographical phenomenon *per se*.

As shown in this chapter, the ROC's changing construction of the ocean corresponds to the theory of territorial political economy. This chapter has demonstrated how persons or organisations intend (or urge others) to use and regulate the ocean can affect how they represent and in turn, construct the ocean. Specifically in this case, after 1949, similar to the state of the Southern Song era facing the military threat of the Jurchen people, the ROC, having withdrawn to Taiwan, confronted the threat from the Chinese Communists. Although mainly protected by the US Seventh Fleet, the ROC maintained a sizeable military force. Largely due to the fiscal burden of the defence of Taiwan, the ROC developed an economic strategy of industrial capitalism; the establishment of a sound seaborne transport system played a crucial role in this development approach.

Reflected by the maritime laws enacted between the early 1950s and late 1980s, the state's construction of the ocean shifted from the Indian Ocean Model towards the Freitian Variation in regulating the maritime activities of Taiwan. After the 1990s, the state began to assimilate key aspects of the Grotian Variation embedded in international maritime treaties in its efforts to transform Taiwan into the APOC in a globalised economy.

Furthermore, the amendments and stipulations of the marine laws and regulations regarding the fishing industry, EEZs and contiguous zones demonstrate that the state's construction of the ocean had absorbed elements of the Seldenian Variation. Moreover, the marine laws regarding the territorial sea showed the incorporation of the components of the Micronesia Model. As the ROC in the 1990s and 2000s attempted to adapt to changing domestic and international economic structures to sustain Taiwan's economic prosperity and to increase the benefits of exploiting maritime resources, its construction of the ocean also evolved, becoming aligned with the construction of the ocean that prevailed in the international treaties of the period. This phenomenon illuminates the theory of territorial political economy.

Steinberg's (2001) theory of territorial political economy is also useful for analysing further changes to the ROC's construction of the ocean after May 2000 when the state, under a DPP-nominated president, endeavoured to cultivate a pro-TIM national identity. The DPP expected that such a national identity reflected the characteristics of an ocean-oriented culture, even though such a culture was only vaguely conceptualised and not clearly elaborated in the official ocean policy white papers.

Taiwanese national identity did not further incorporate the ocean into its nationalistic discourses during the KMT administration. Instead, the state's focus on the sea shifted from the ocean's ideological, discursive role to its economic development function. Nevertheless, in the following chapters, I shall examine how the economic, military and ideological sources of power in modern Taiwan society constructed the ocean. I first deal with the construction of the ocean generated by the sources of ideological power.

Notes

1 This chapter is a modification and expansion of my journal paper, see Hou (2014).
2 Those important questions about stipulating and implementing marine policies have been addressed at the beginning of this monograph's first chapter.
3 In China, the pre-modern era means the long historical epoch before the mid-nineteenth century.
4 Hereafter, I use the term, 'the 1958 Maritime Act', to refer to the Maritime Act amended in 1958 and the term, 'the 1929 Maritime Act', to denote the original version of this act. Notably, in this monograph, the number assigned before the name of an act means the year in which the act was first stipulated or later amended.

References

Chang, Yu-fa (2001) *Zhonghua Minguo Shigao*, xiuding ban (A Draft of the History of the Republic of China, revisited edition) (Taipei; Linking).
Cole, Bernard (2010) *The Great Wall at Sea*, 2nd edition (Annapolis; Naval Institute Press).
Fu, Kuen-chen (2001) *United Nations Convention on the Law of the Sea with All Annexes and Related Agreements and a Chinese-English Index*, 2nd edition (Taipei; Xierong).
Hou, Kuang-hao (2014) 'The Meaning of the Ocean for the State in Contemporary Taiwan since 1949: Understanding State Perception of the Ocean through a Historical Approach', *Australian Journal of Maritime and Ocean Affairs*, 6(4), 191–206.
Legislative Yuan (1958) *Lifayuan Gongbao* (Legislative Yuan Gazette), Volume 47, Number 21(5).
Legislative Yuan (1962) *Lifayuan Gongbao* (Legislative Yuan Gazette), Volume 59, Number 29(13).
Legislative Yuan (1968) *Lifayuan Gongbao* (Legislative Yuan Gazette), Volume 57, Number 7.
Legislative Yuan (1969a) *Lifayuan Gongbao* (Legislative Yuan Gazette), Volume 58, Number 21.
Legislative Yuan (1969b) *Lifayuan Gongbao* (Legislative Yuan Gazette), Volume 58, Number 34.
Legislative Yuan (1981) *Lifayuan Gongbao* (Legislative Yuan Gazette), Volume 70, Number 15, 10.
Legislative Yuan (1991) *Lifayuan Gongbao* (Legislative Yuan Gazette), Volume 80, Number 25.
Legislative Yuan, (1996a) *Lifayuan Gongbao* (Legislative Yuan Gazette), Volume 85, Number 29.
Legislative Yuan (1996b) *Lifayuan Gongbao* (Legislative Yuan Gazette), Volume 85, Number 37.
Legislative Yuan (1997) *Lifayuan Gongbao* (Legislative Yuan Gazette), Volume 86, Number 11.
Legislative Yuan (no date 1) *Lifayuan Falü Xitong* (The Laws System of the Legislative Yuan), Lifayuan Ziliaoku (Database Retrieval Services), 'Hangye Fa' (The Shipping Act) <http://lis.ly.gov.tw/lgcgi/lglaw>.
Legislative Yuan (no date 2) *Lifayuan Falü Xitong* (The Laws System of the Legislative Yuan), Lifayuan Ziliaoku (Database Retrieval Services), 'Haishang Fa' (The Maritime Act) <http://lis.ly.gov.tw/lgcgi/lglaw>.

Legislative Yuan (no date 3) *Lifayuan Falü Xitong* (The Laws System of the Legislative Yuan), Lifayuan Ziliaoku (Database Retrieval Services), 'Yuye Fa' (The Fishery Act) <http://lis.ly.gov.tw/lgcgi/lglaw>.

Lin, Chin-bi (2007) *Haishangfa Lun* (On the Maritime Law) (Taipei; Sanmin).

Lo, Jung-pang (2012) *China as a Sea Power, 1127–1368: A Preliminary Survey of the Maritime Expansion and Naval Exploits of the Chinese People during the Southern Song and Yuan Periods*, edited, and with commentary, by Bruce Elleman (Hong Kong; Hong Kong University).

Mann, Michael (2012) *The Sources of Social Power*, Volume 2: The Rise of Classes and Nation-states, 1760–1914, new edition (Cambridge; Cambridge University Press).

Migdal, Joel (2001) *States in Society: Studying How States and Societies Transform and Constitute One Another* (Cambridge; Cambridge University Press).

Ministry of Interior (2002) *Haiyu Erfa Zhiding Shilu* (The Record of Stipulating the Two Marine-territorial Laws) (Taipei; Ministry of the Interior).

Steinberg, Philip E. (2001) *The Social Construction of the Ocean* (Cambridge; Cambridge University Press).

Wade, Robert (2004) *Governing the Market: Economic Theory and the Role of Government in East Asian Industrialization*, 2nd edition (Princeton; Princeton University Press).

Weale, Albert (2004) 'Politics as Collective Choice', in Adrian Leftwich (ed.), *What Is Politics?* (Cambridge; Polity Press), 86–99.

Yin, Chang-hua (2003) *Guoji Haiyangfa* (The International Law of the Sea) (Taipei; Wensheng).

Zheng, Yangwen (2012) *China on the Sea: How the Maritime World Shaped Modern China* (Leiden; Brill).

3 The Social Constructions of the Ocean by the Ideological Sources of Power in Modern Taiwan[1]

Introduction

In this chapter, I aim to demonstrate that an intellectual's social construction of the ocean can be shaped, if not completely determined, by how their intention to use the ocean or by how they urge others to use the ocean. This argument corresponds to Philip E. Steinberg's (2001) analysis and theory. In addition, demonstrating this argument presents a picture of how the intellectuals on the modern island of Taiwan socially constructed the ocean. The previous chapter answers the question as to how the Republic of China—the state now governing Taiwan, Kinmen and Matsu constructed the ocean between 1949 and 2016. I have suggested this due to historical legacy and the promotion of economic development—the state in the society of modern Taiwan formulated a hybrid construction of the ocean, mixing different attributes of the three models of the Steinbergist Classification in varying proportions.

The elements that were mixed are closely related to the elements of the ocean that the state intended to utilise. But how did the other sources of power in the society of modern Taiwan construct the ocean? The analysis presented in the second chapter is a part of my attempt to present a picture of multiple and diversified constructions of the ocean in modern Taiwan. This picture helps us understand the human-marine relationships and the politics of marine policy stipulation and implementation in Taiwan. Subsequently, I present a piece of the jigsaw puzzle that this monograph is trying to complete by studying the intellectuals' written works in modern Taiwan.

Human societies consist of sources of social power (Mann 2012a; 2012b). Among these sources, the four most important types can be identified as state, ideological, military and economic sources. How the state governing Taiwan and Kinmen and Matsu constructed the ocean has been analysed in the previous chapter. This third chapter analyses the social constructions of the ocean generated by the ideological sources of power in the society of modern Taiwan.

As Michael Mann (2012a; 2012b) identifies, ideological sources emerge as a consequence of the human need for concepts and theories (for example,

DOI: 10.4324/9781003291343-3

structural functionalism) to comprehend their world, norms and values (for example, Confucianism) to regulate actions, and aestheticism and religion (for example, Christianity) to find the ultimate meanings in life. It is apparent that intellectuals in Taiwan (for example, academics, literary writers and historians) are society's primary constituents of the ideological sources of social power.

The intellectuals' works are divided into two parts in this chapter. For the first part, I employ the Steinbergist Classification to examine and interpret the natural and social scientists' works within modern Taiwan—regarding maritime and ocean affairs. These scholars were considered prominent and influential because they greatly contributed to the relevant studies in the society of modern Taiwan. Their writings demonstrate how they represented the ocean, how they advised the use of the ocean, and how they advocated what human-marine relations should be. Consequently, their writings show how they constructed the ocean.

For the second part, I analyse the works of important intellectuals in the fields of literary and history, relevant to maritime and ocean affairs. I examine what human-marine relations these intellectuals advocated to build and evaluate the utility that these intellectuals' writings contributed to the formation of Taiwanese nationalism and Taiwanese national identity.

In the case of the first part of the intellectuals' works, scholars urging the state and society on Taiwan to exploit the resources of the ocean tended to construct the ocean according to the Micronesia Model. In contrast, intellectuals who wrote about how people in Taiwan were utilising the surface of the ocean without analysing the ocean per se, and without advocating the exploitation of marine resources, were constructed the ocean according to the Mediterranean Sea Model.

In the case of the second part of the intellectuals' works the historians' works tend to focus on maritime trade and migration and view the ocean as a great void. These scholars constructed the ocean according to the Indian Ocean Model. Moreover, the literary works highlight the same tendency (with Syaman Rapongan's literature as an exception). I argue that the construction of the ocean, presented in the literary writers' and historian's works evaluated here, cannot provide useful ingredients for adding new elements into Taiwanese nationalism and national identity. This is because their social constructions of the ocean are consistent with the Indian Ocean Model, as is the construction of the ocean generally held by traditional Chinese states and societal actors. The exception of Syaman Rapongan's construction did not change this situation because his works are serving a distinct aboriginal ethnic identity. This was alienated from the Taiwanese *qua* Han Chinese ethnic and national identities, at least during the time of observation by this study.

This chapter argues that how people intend, and urge others, to use the ocean affects how they construct it. This perspective is particularly evident through the works of Taiwan's natural and social scientific intellectuals of maritime and ocean affairs. It echoes and complements one of the primary theses of Steinberg's (2001) theory of territorial political economy; that is, the use, regulation and representation of a geographical phenomenon are interlinked to, and influenced by each other, effectively creating a social construction of the phenomenon.

The Social Construction of the Ocean by Natural Scientists in the Society of Modern Taiwan

Human beings require concepts and theories to help them explain and interpret the world. Concepts and theories are required in social sciences and necessary for people to acquire knowledge about nature. Modern natural scientists have compiled numerous concepts and theories about the natural world, which affect human material well-being. Thus, they constitute a crucial part of the ideological sources in the society of Taiwan. Therefore, this chapter examines the relevant literature on maritime and ocean affairs composed by natural scientists.

In this chapter, I suggest that the natural scientists' constructions of the ocean belonged to the Micronesia Model because their writings, first, reveal the plentiful and abundant resources that the ocean can provide and, second, promote further utilisation and sustainable development of the resources. The objectives and purposes of their studies and writings shape their constructions of the ocean.

The development of natural scientific knowledge of Taiwan's academia about the ocean began when natural scientists employed knowledge of their original disciplines that had been previously used to study natural phenomena on the land (for example, thermodynamics or biochemistry) to explore the ocean (Hu 1997). Therefore, according to their writings, the ocean means more than a neutral geographical phenomenon. It becomes a depository and reservoir of knowledge waiting to be pumped. The ocean offers knowledge, which can be viewed as a kind of immaterial resources. Moreover, the more scientists learned about the ocean, the more they realised further opportunities to utilise the ocean. Therefore, through their literature, the ocean is constructed as a resource reservoir as in the Micronesia Model.

Academic works written by natural scientists can be difficult for a layman to understand. However, their works, composed for the general public, such as elementary oceanographic books of popular science or policy advisory papers, are accessible. These forms of writing typically use more words and fewer formulae to convey ideas to readers. In effect, these works reveal more about their constructions of the ocean than their technical and professional counterparts.

The main argument is supported by the evidence derived from the policy advisory papers collected in the three editions of *Haiyang Taiwan: Yongxu Fazhan* (Oceanic Taiwan: Sustainable Development) published respectively in 2003, 2009 and 2014.[2] These papers were first submitted to a series of the largest and most important national marine policy conferences on Taiwan prior to publication. The majority of important scholars in modern Taiwan, whose works are related to the ocean and maritime affairs, attended these conferences.

The natural scientific advisory papers on marine policy, in the three editions of Haiyang Taiwan, contain various topics. These topics are: (a) the protection of coastal areas; (b) the preservation and sustainable development of marine resources; (c) the protection of marine biodiversity; (d) the development of marine energy; (e) further explorations of the deep sea and the development of technology for oceanic explorations; and (f) managing the challenges of climate change. This section focuses on the topics (a), (c) and (e) as examples to demonstrate the aforementioned primary argument.

With respect to topic (a), the papers include Lee *et al.* (2003) and Lee, Kong, and Yang (2009). These two papers were co-authored by marine engineering scholars. The first paper deals with the causes of the erosion of coastal areas, problems that require attention in the coastal areas and possible solutions for the present problems and strategies of protection for the areas from future damage. The second paper begins with the current uses of the coastal areas of Taiwan, discusses the current problems for these areas and finally offers suitable approaches for the protection of the areas. In addition, the current governmental methods of protecting Taiwan's coasts from erosion by the ocean are evaluated.

The construction of the ocean by these two papers is not one of the Micronesia Model. Although they promote approaches to protect Taiwan's coastal areas, the ocean is reduced to a threat. The ocean is perceived as an empty space, and its fringe can expand. It can damage and erode the land. In this construction, humans and states in particular, must find solutions to tame the encroaching ocean and defend the coastal land. Efforts must be made by the state to govern and manage the negative impact of the ocean. This is a construction similar to the original Mediterranean Sea Model.

With respect to topic (c), Kwang-tsao Shao (2003; 2009; 2014) wrote one paper for each edition of Haiyang Taiwan. Shao (2003) begins with a brief introduction of how many types of marine animals have been recorded and examined in the ocean surrounding Taiwan. More than 2,500 species of fish, and hundreds of other forms of ocean animals, have been registered—further elaborating on the crises facing the marine eco-systems of Taiwan. Finally, it presents feasible policy advice for the state to protect the marine eco-system.

Shao (2009) consistently follows the rationale of the policy advices offered in his previous paper. He highlights that Taiwan's marine environment contains numerous species and that the level of biodiversity is very high. Moreover, the system of marine habitats on Taiwan's offshore areas is comprehensive and complete. However, he urges the state to establish sensitive zones (or hotspots) of the marine eco-systems to protect marine biodiversity within Taiwan's sea areas.

Shao (2014) provides updated information on Taiwan's marine eco-systems and promotes the notion and policy of establishing protected marine areas.

In addition, he compares other countries' protected marine areas with those of Taiwan in order to highlight what more can be done by Taiwan's government to protect its marine biodiversity.

Employing the Steinbergist Classification as an observational lens, I suggest that the commonalities among Shao's (2003; 2009; 2014) papers are not only their request for the state to protect marine biodiversity on Taiwan's surrounding ocean areas, but also their representation of the ocean as a great source of various forms of marine life. This marine life constitutes precious material and non-material resources, for example, ingredients of medicine and knowledge of special life forms. The ocean in these papers has been constructed as a resource reservoir, which is a notion of the Micronesia Model.

With respect to topic (e), the papers include Hwung (2003); Chiu (2009) and Jan, Hsieh and Dai (2014). In the three editions of Haiyang Taiwan, Hwung (2003) is the first to demand that the state must launch a full-scale investigation into the marine environment and establish a comprehensive database of the seas surrounding Taiwan. Chiu (2009) elaborates on the advantages of employing a high-frequency radar system for exploring the ocean. Jan, Hsieh and Dai (2014) discuss the benefits of deploying special ships for marine research and creating an integrated database on marine sciences and environment.

Hwung (2003), as a scholar of hydraulics, identifies the plentiful resources on Taiwan's surrounding seas and convincingly suggests that sustainable utilisation and development of these resources depend on comprehensive investigations and on establishing well-organised exploration and well-informed policies. The use of the radar systems can help to achieve this.

Finally, three oceanographers, Jan, Hsieh and Dai (2014), offer the policy advice and introduce the current status of information integration in the fields of marine physics, marine biology and marine chemistry. Their introduction highlights the complexity of the ocean and can provide both material resources (for example, minerals and fish) and non-material resources (for example, knowledge and recreation). The papers of this third topic construct the ocean as a resource reservoir, as does the Micronesia Model.

To summarise this section with a comparison of the papers previously discussed, those on topics (c) and (e) have identified a larger scope of resources than mere fisheries that the ocean can provide, whereas the papers on topic (a) have focused not on the ocean, but on how to prevent the erosion of the land by the ocean. The papers discussed in this section demonstrate that the intentions of the natural scientists' written works affect the constructions of the ocean in their work. Natural scientists' writing promoting the utilisation and preservation of marine resources (including knowledge about the ocean) are inclined to construct the ocean according to the Micronesia Model. Effectively, this supports how the natural scientists' aims of using, and of urging people to use, the ocean in their writings influence how they represent and construct the ocean.

The Social Constructions of the Ocean by Economic and Managerial Academics in the Society of Modern Taiwan

The social scientists on modern Taiwan (before 2016) who wrote about marine-related issues were primary economic and managerial academics, in addition to being political and legal scholars. This section introduces the work of several economic and managerial academics who were well-known for their academic studies on maritime and ocean affairs. The following section analyses the works of the political and legal intellectuals.

The selection of the intellectuals is not based on random sampling but on their reputations and importance. They were employing their knowledge of conventional disciplines to explore maritime and ocean affairs. Subsequently, they were influential in their academic contributions to these affairs.

Why the intellectuals chose to write about a certain topic could be personal and contingent. However, because the mainstream economics and managerial sciences aim to minimise costs, maximise profits and maintain the competitiveness of an organisation in the processes of the extraction, transformation, distribution and consumption of resources, these intellectuals often concentrated on maritime and ocean industries, including seaborne transportation, ports, fisheries, marine minerals and oceanic tourism.

This section begins with LIN Kuang's papers and textbook. The former vice-president of the National Taiwan Ocean University, Lin (a successful entrepreneur), established the Pacific Star Group, the largest international freight forwarder company in Taiwan, in 1977. Being an influential entrepreneur-scholar, he is friends with many university academics in the field of maritime shipping management in Taiwan. The information on Lin's personal achievements shows how he uses the ocean.

There are several characteristics of Lin's academic papers. First, these papers concentrate on issues of maritime shipping and port management. For example, the international trends of maritime shipping (Lin and Ni 1998), the relationships between the freight forwarding industry and business networks (Lin *et al.* 1999), and the evaluation of the performance of international harbours (Liang, Lin and Chou 2003).

Second, these papers contain technical jargon and are primarily concerned with increasing the competitive advantages of maritime shipping and port industries. The ocean per se is not discussed in these papers. For example, Lin *et al.* (1999) discuss utilising business networks to improve the competitiveness of maritime freight forwarders. The paper uses many tables and charts to explicate the structures and patterns of the processes of employing business networking in the freight forwarding industry. Similarly, the paper written by Lin, Lee and Wang (2003) evaluates the financial performance of the maritime shipping industry on Taiwan.

Third, LIN Kuang provided advice for governmental decision-makers several papers in order to increase the profit of the maritime shipping industry. For example, he promoted using larger ships capable of carrying more, larger containers and building deeper harbours to receive these larger ships (Lin and Ni 1998). He presented a list of methods for making harbours on Taiwan more efficient (Lin and Wang 2000).

Regarding his textbook written for university students, he wrote the first edition of *Haiyunxue* (Maritime Shipping) in 1983. The second edition, published in 1985, and the ninth edition, published in 2012, are examined in this section. By comparing these 2 editions, it is evident that the way this textbook constructs the ocean has not dramatically changed for 30 years.

The textbook reveals how the author represents and constructs the ocean. The list of topics included in the second edition is long but worth noting here. It reveals how the author represents and constructs the ocean: the development of the shipping industry in modern China; the contemporary conditions of the shipping industry in the world; shipping and sea power; ships; the tonnages of ships; cargoes; commercial harbours; captains and seafarers; berth services and tramp services; bills of lading; and leasing of vessels and charter parties.

The list continues as follows: the shipping industry; freight forwarders; loading brokers; customs brokers; consolidators; non-vessel operating common carriers; shipping agencies; flags of convenience; the market and rates of shipping; shipping conferences; marine insurance; protection and indemnity; the organisation of shipping companies; the Shipping Act and Maritime Act; the Hague Rules and Hamburg Rules; maritime documents; container transportation; and maritime policy.

The long list of primary topics remains relatively unchanged since the textbook was first published. Majority of these topics are retained in the latest edition, although the textbook removes the discussion of sea power and later editions have been supplemented with newer information and academic findings. In addition, the second edition of the textbook emphasises the importance of maritime shipping if the state decides to launch military campaigns against the Chinese Communists across the Taiwan Strait. Moreover, it identifies that experienced seafarers are the reserved sailors of the navy. Furthermore, it indicates the functions that maritime shipping can serve for stimulating economic growth in sectors such as ship building, harbour construction, and integrating and promoting terrestrial transportation. However, in the ninth edition, the textbook removes their discussion of sea power but still stresses the importance of maritime shipping for the economy and culture of Taiwan (Lin, Chang and Chao 2012).

LIN Kuang's academic works (and his use of the ocean through his freight forwarding company) have constructed the seas in the style of the original version of the Mediterranean Sea Model in his early career, and, later, in the manner of the Freitian Variation of the Mediterranean Sea Model. First, his earlier writings, especially in the second edition of the textbook, advocate state governance of maritime shipping, and highlight the utilities of a sound seaborne transportation system for the Republic of China's (ROC's) sea and naval powers and Taiwan's economic vitality. This is a construction pertinent to the original Mediterranean Sea Model. The long list of the topics of the textbook is unrelated to the ocean resources, nor do the topics discuss the ocean itself.

Nevertheless, the construction has changed. The ninth edition of the textbook is almost completely free of a discussion of the benefits of seaborne transportation for military operations. What remains is a short section noting the contributions of maritime shipping to national defence. Moreover, his papers––published since the late 1990s—have demonstrated greater interest in profit maximisation. Therefore, his original Mediterranean construction of the ocean has changed into one of the Mediterranean-Freitian type, which stresses the state's responsibility to increase shipping industry profits. The way in which Lin's writings have constructed the ocean sheds highlights the following viewpoint: how a person intends, or urges others, to use the ocean can influence their representation of the ocean. Subsequently, this can shape the second nature of the ocean.

This section analyses the work of CHUANG Ching-ta, a marine economist, a crucial academic staff member and a former director of the Institute of Marine Affairs and Resources Management (National Taiwan Ocean University)—one of the most important graduate institutions of maritime and ocean affairs. His numerous publications are influential.

Chuang began his career as an economist working in the agriculture and fishery fields. His earlier works deal with several topics, however, after the mid-1990s he concentrated on ocean-related issues. His writings construct the ocean in two different ways. His works about Taiwan's fishing sector construct the ocean in the way of the Grotian Variation, whereas his publications about the governance of oceanic tourism and marine resources construct the ocean according to the Micronesia Model.

Chuang's works on the fishing industries on Taiwan (published after the mid-1990s) indicate that he began to focus on the state's role in helping the fishing sector to survive, transform and thrive. His works suggest that the state should form a fishing alliance with Japan and actively govern and regulate fishing activities in order to prevent the over-exploitation of fish stocks. In addition, his publications suggest that the state must improve the welfare of fishermen, facilitate the sustainable development of fishing villages and alleviate the negative influences of Taiwan's participation in the World Trade Organization (WTO) on the fishing sector (Chuang, Ho and Liao 2015; Chuang, Liu and Wu 1998).

As a consequence of Taiwan joining the WTO in 2002, Chuang (2011) suggests that the fishery policy of the state must be adjusted in order to guide Taiwan's fishing industries and maintain its competitiveness in the globalised market. Moreover, the state must manage the fishery zones where Taiwan's fishermen work to make the fishing industries obey the international regulations and principles of the sustainable utilisation of fishing resources.

Notably, the ocean *per se* is not mentioned in those papers. However, he emphasised the responsibility of the state to regulate fishing activities.

Moreover, his advice on the establishment of a fishing alliance with Japan was a positive response to the international fishery regime: the institution of responsible fishing. Therefore, his works on the fishery on Taiwan contain a social construction of the ocean pertinent to the Grotian Variation. His works also urge the state to foster oceanic tourism to help the fishing industries cope with the impact of Taiwan joining the WTO.

His written works on oceanic tourism and marine resources represent the ocean as a resource reservoir and thus construct the ocean through the Micronesia Model. In a paper examining environmental conservation on Taiwan's Three Northern Islets Marine National Park (Jan, Cheng and Chuang 2011), marine animals and special marine topographical characteristics in the national park are mentioned, and the notions of the terrestrial sea and the marine protection area form the background of the discussion in the paper. In a paper examining the development of ornamental fisheries on Taiwan (Chen, Chuang and Kung 2012), the ocean is described as a major source of fish for aquarium purposes. The ROC is expected to properly manage its own ocean zones where ornamental fishes are produced.

In a nutshell, the above two papers present the ocean as a resource reservoir and then construct the ocean according to the Micronesia Model. However, in order to better understand the representation and construction in his works on oceanic tourism (including the tourist fishery), two of his textbooks on marine resource governance require examination: *Haiyang Ziyuan Guanli: Lilun yu Shiwu* (Marine Resources Management) and *Haiyang Shiwu Gailun* (Introduction to Marine Affairs).

In his written works, marine resources are broadly defined as the elements of the ocean that can be exploited and employed by humans, while oceanic tourism is deemed as a form of marine resource. Marine resources include 'marine minerals, sailing routes and harbours, marine energy, marine chemical resources, and marine living resources' (Chuang *et al.* 2008: 15). Humans must endeavour to develop the ocean in order to extract the elements of the ocean and transform them into utilities to benefit humans (Chuang *et al.* 2013).

Chuang's textbook (Chuang *et al.* 2008) identifies several characteristics of the development of marine resources. They highlight the book's representation and construction of the ocean. First, increasingly more states are viewing the development of marine resources as their national strategic goal. Second, the development of the ocean involves multiple objectives and plural interests. It requires the state's co-ordination and meticulous planning. Third, the development of marine resources requires scientific governance in order to achieve the sustainable development of the ocean (Chuang *et al.* 2008).

Fourth, the development of marine resources has been a nascent phenomenon. For example, the history of tapping marine energy only began around 20–30 years ago. Fifth, knowledge about the development of marine resources is inter-disciplinary. Making a sound policy of ocean development demands an adequate understanding of inter-disciplinary knowledge. Sixth, the development of marine resources is related to international politics.

The use of the ocean can easily create conflicts of interest between states. Seventh, the development of marine resources inevitably impacts, and is unavoidably shaped, by the reactions of the ocean. Eighth, the abundant resources of the ocean can become the necessary substances for human survival (Chuang *et al.* 2008).

In addition to the description of the ocean as a resource provider, Chuang's textbooks also introduce important international maritime regimes, such as the International Convention for the Safety of Life at Sea and the International Maritime Organization. However, his introduction of international maritime regimes, navigation, fishery and other marine resources is entirely based on the United Nations Convention on the Law of the Sea (UNCLOS). Moreover, reasonable and responsible utilisation and development of the marine resources in a state's territorial seas, and other demarcated marine zones granted by UNCLOS, are highlighted in the textbook (Chuang *et al.* 2013).

As demonstrated, the ocean becomes a claimable territory and a possessable resource reservoir. The ocean and its resources can (and should be) owned, regulated and governed by the pertinent state. Therefore, the works mentioned earlier on the marine resources construct the ocean in the Micronesian style.

Scholars who focused on maritime shipping tended to construct the ocean as a surface on which the state was expected to govern important routes of seaborne transportation. This was a construction of the ocean with Mediterranean-Freitian characteristics. Scholars who concentrated on contemporary fisheries were inclined to construct the ocean as a space arena subject to the rules of international regimes. This was the construction of the ocean in a Mediterranean-Grotian style. Moreover, scholars who produced works on marine resource governance constructed the ocean according to the Micronesia Model.

An author's different works analysing and introducing various uses and regulations of the ocean can construct the ocean in diversified ways. Similarly, scholars in the same discipline (or closely related ones) could construct the ocean in distinctive manners when their works analysed special utilisations of the ocean. Thus, the overarching implication is how people intend, or urge others, to use the ocean influences how they represent the ocean, and hence, shapes the second nature (or the social construction) of the ocean.

The Social Constructions of the Ocean by Legal and Political Scholars in the Society of Modern Taiwan

In this section, I explicate the social constructions of the ocean embedded in the writings of legal and political scholars in the society of modern Taiwan primarily between the mid-1990s and 2016. When the legal and political scholars wrote about issues of maritime and ocean affairs, they shared two commonalities. First, both groups of intellectuals constructed the ocean as a claimable territory and a possessable resource reservoir in their written works. Second, UNCLOS was often the main reference in their writings. This section demonstrates evidence supporting the main argument of this chapter: how intellectuals intend, or urge others, to use the ocean in their writing shapes their representations of the ocean and thus influences their constructions of the ocean.

The legal scholars' written works on marine-related issues are primarily set in the context of public international law, and UNCLOS forms the framework of their discussions. For example, UNCLOS forms the backbone of *Guoji Haiyang Fa* (The International Law of the Sea), written by Huang-chih Chiang (2004), and which is the most comprehensive and important textbook on the subject available in the Chinese language in modern Taiwan society. In addition, legal scholars' works on marine issues often examine disputes over the delimitations of marine areas in the East and South China Seas. Their arguments are anchored on UNCLOS and related conventions.

Chiang's (2013) paper examines the legal status of the Okinotori coral reefs. While this paper convincingly suggests that the Okinotori should be deemed legally as reefs in the East China Sea, it criticises the ambiguous definition of the island provided by UNCLOS—which is the framework of analysis in the paper. Similarly, Yann-huei Song's (2015) paper focuses on Taiping Island and attempts to answer the questions about whether Taiping is an island or rock under UNCLOS. Importantly, their reference to UNCLOS constitutes an attempt to demarcate the ocean zones and assists the state on Taiwan to claim and possess its legally granted sea areas.

Where the legal intellectuals discuss more domestic-oriented marine issues (for example, the protection of the underwater cultural heritage around Taiwan Island, or the stipulation of the two marine laws on Taiwan), UNCLOS still constitutes the context of the discussions. *Haiyang Guanli de Falü Wenti* (Ocean Management and Law), written by an authoritative expert on marine laws, Kuen-chen Fu (2003), exemplifies this academic trend.

When modern Taiwan's political scholars attempted to evaluate Taiwan's marine affairs, the majority of them also analysed the disputes over maritime delimitations in the East and South China Sea. However, they were inclined to place these disputes in the wider context of international power struggles amongst several strong states, such as Sino-American or Sino-Japanese relations.

Undeniably, the writings of the political scholars who studied sea power constructed the ocean in the fashion of the original Mediterranean Sea Model, because they concentrated on naval strength, war strategies, battle tactics, and military and geopolitical goals (Chang 2013). The political scholars of sea power in modern Taiwan wrote about the ocean, representing the seas as an arena where states competed for military domination. Their works provided evidence supporting the main argument of this chapter.

However, for scholars of diplomacy and international politics, UNCLOS and the exclusive marine area are still the conceptual basis for their discussions on marine area delimitations in the East and South China Seas.

Consequently, their works contend that these disputes and confrontations over oceanic demarcations are closely related to, if not completely caused by, the norm and practice of appropriating portions of the ocean.

Several political scholars in Taiwan, who wrote about marine affairs without belligerent intonations, analysed the issues of international regimes and the domestic policies of ocean governance. For example, Kien-hong Yu (2015) contributed a book titled *Ocean Governance, Regimes, and the South China Sea Issues*. It criticises UNCLOS from the perspective of international regimes. He argues that one of the deficiencies of UNCLOS is its failure to function as a coherent international regime. For him, a functioning international regime is meant to resolve common problems between the concerned parties. The internal inconsistency of UNCLOS is deemed as a crucial origin of the clashing territorial claims in the South China Sea.

Regarding the writings on the domestic policies of ocean governance in contemporary Taiwan society, those composed by Nien-tsu Hu (a well-known marine policy researcher and the director of the reputable Graduate Institute of Marine Affairs, National Sun Yat-sen University) and Wen-yan Chiau (a proactive marine policy researcher and a former member of Taiwan's central legislative branch, the Legislative Yuan) are examined here. Generally, the former emphasises the necessity of a ministry of marine affairs and the latter stresses the sustainable development of Taiwan's coastal areas and other marine resources. They both believe that Taiwan's marine environment and resources can benefit from a strong governmental organisation capable of co-ordinating the maritime and ocean affairs in modern Taiwan.

Discussing their relevant works allows us to understand their constructions of the ocean, which support the contention of this chapter. In Hu's (1997) earlier book, having compared marine policies in other countries, he suggests that the ROC should establish an inter-ministerial committee to co-ordinate the implementation of marine policies. His reasons for building such a committee are listed below—they show how he has constructed the ocean's second nature in that book.

First, marine, coastal and terrestrial areas constitute the territories of the state. Not only the land, but also the ocean and coast can be possessed and require the state's governance. Second, in order to effectively use marine resources, an integrated institutional arrangement should be established, because an adequate utilisation of marine resources is a complex matter that requires trans-departmental co-ordinations. For example, the development of marine energy needs co-operation between the governmental agencies of (at least) the fishery and energy management.

Third, oceanic law enforcement is necessary, because UNCLOS has clearly granted rights for coastal states to utilise, protect and govern their legally demarcated sea areas. In order to facilitate adequate law enforcement on the ocean, a co-ordinating central governmental organisation has to be installed. Fourth, UNCLOS and its relevant conventions often require states to comply with regulations for achieving the sustainable utilisation and development of fisheries and other marine resources. Therefore, the ROC should establish a central governmental organisation responsible for dealing with these requirements and for co-ordinating actions among the concerned governmental branches.

In his paper about the institutional design of the ministry of marine affairs, Hu (2002) mentions the reasons listed earlier and identifies the problems that a mere co-ordinating inter-ministerial committee of marine affairs can encounter. He subsequently suggests that the co-ordinating committee will have insufficient authority to overcome bureaucratic resistance in other related ministries that actually require co-ordination. Nevertheless, his reasons for establishing the central governmental organisation of marine affairs demonstrate that his works have constructed the ocean according to the Micronesia Model. Following UNCLOS, his written works construct the ocean as a possessable resource reservoir.

Similarly, Chiau (2014) also deems it necessary to form a central governmental organisation to co-ordinate the implementation of marine policies in Taiwan, although he accepts establishing an inter-ministerial committee as the first step towards the gradual creation of a more comprehensive and powerful ministry of marine affairs (Chiau 2014: 53). His written works discuss marine policies (and the design of the central governmental organisation of marine affairs) and introduce coastal management. Chiau (2001) supports the establishment of a strong central ministerial organisation of marine affairs because of his promotion of integrated coastal management and his endorsement of the sustainable development of Taiwan's marine and coastal resources.

Integrated coastal management perceives the ocean as an extension of the land and deems it necessary to manage coastal and oceanic area as a whole. For Chiau (2001), integrated coastal management is an interactive process between humans and the oceanic and coastal areas to sustainably utilise marine and coastal resources. The aims of integrated coastal management include:

> first, to maintain the integrity of the resource systems of the coastal areas; second, to reduce the conflicts caused by the utilisation of oceanic and coastal resources; third, to maintain the health of the environment; and fourth, to facilitate the development of oceanic and coastal resources.
>
> (Chiau 2001: 7)

Moreover, for a state to undertake integrated coastal management, it must achieve 'the integrations, first, of the different departments responsible for governing various oceanic and coastal resources (for example, the fishery and oceanic tourism); second, between the aquatic and terrestrial zones within the parameter of the coasts; third, among the central and local governments; fourth, among states; and fifth, among different spheres' (for example, natural and social sciences; Chiau 2001: 7).

Chiau (2001) perceives integrated coastal management (and an institutionalised inter-ministerial committee) as a pathway to sustainable development.

He also believes that it corresponds to the request of UNCLOS for each coastal state to control, use and conserve its oceanic and coastal areas and resources. Although two of Chiau's (2005; 2007) important academic papers introduce information about the resources on Taiwan's marine and coastal areas, these two papers follow the requests of UNCLOS and the ideals of sustainable development—rather than the rationale of profit maximisation. Moreover, in his 2007 paper, he embraces the term 'marine territorial planning' and explicates its relevant policies implemented in other countries.

Chiau's works have identified that the oceanic zone and coastal area are an inseparable whole and the continuation of the land; indicated that oceans, coasts and lands can all be possessed and become a state's territory; and convincingly suggested that a central governmental organisation should be established to adopt the integrated approach to the sustainable development of Taiwan's oceanic and coastal resources (Chiau 2014). Therefore, it can be argued that his written works have constructed the ocean according to the Micronesia Model.

To summarise this section, legal and political scholars shared two commonalities in their construction of the ocean. First, these two groups of intellectuals constructed the ocean as a claimable and possessable territorial resource reservoir. Second, UNCLOS often becomes the backbone of their discussions on maritime and ocean affairs. The works of political and legal scholars' have constructed the ocean according to the Micronesia Model.

Furthermore, their works have either implied the potential utilisation or advocated the active development of abundant marine resources. Moreover, several of them have urged the state to conserve its oceanic and coastal resources by establishing a central ministerial organisation of marine affairs. On the contrary, the writings about naval power have constructed the ocean in terms of the original Mediterranean Sea Model, in which they have urged the use of the ocean for military purposes and the enhancement of the state's sea power.

The discussion regarding legal and political scholars' writings of maritime and ocean affairs in this section has demonstrated the view of how people intend to use, or urge others to use, the ocean in their written works can shape how they represent this geographical phenomenon and its second nature.

The Social Constructions of the Ocean by Intellectuals in the Field of Humanities in the Society of Modern Taiwan

The sections presented earlier constitute the first part of this chapter. I will now proceed to the second part and discuss the social constructions of the ocean by the humanities intellectuals (mainly literary writers and historians in Taiwan primarily between the 1980s and 2016).

To begin with, the historians studying topics related to maritime and ocean affairs generally focused on maritime and migratory history. Their constructions of the ocean followed the Indian Ocean Model—the traditional Chinese states' way of constructing the ocean as mentioned in Chapter 2. Furthermore, an oceanic literary writer, LIAO Hung-chi, also constructed the ocean in the style of the Indian Ocean Model. Although Syaman Rapongan did construct the ocean according to the Micronesia Model from which the state ruled by the Democratic Progressive Party (DDP) would attempt to extract elements for enhancing the Taiwanese national identity, his representation of the ocean was also his discourse aiming to develop a distinct ethnic identity for the indigenous people on Orchid Island (Lanyu). Syaman Rapongan's construction of the ocean could not help the state reigned over by the DPP to accomplish the task of Taiwanese nation-building.

This section demonstrates the primary argument of this chapter and the theory of territorial political economy. The scholars in the humanities interpreted how others used the ocean. Effectively, the uses of the ocean—studied and described by these intellectuals—could affect the intellectuals' representations and constructions of the ocean. The majority of humanities scholars in modern Taiwan, during the period examined in this section (except Syaman Rapongan) constructed the ocean in the manner of the Indian Ocean Model. Syaman Rapongan described and interpreted how his tribal fellows' lives were integrated with the ocean. In accordance with Steinberg's (2001) theory of territorial political economy, Syaman Rapongan's construction of the ocean belonged to the Micronesia Model.

Historical Works of Maritime and Migratory Activities

One of the most prominent historical studies on modern Taiwan, regarding maritime and migratory activities, is *Zhongguo Haiyang Fazhan Shi Lunwen Ji* (Essays in Chinese Maritime History). This collection of papers on maritime and migratory history consists of ten volumes. The publication of these volumes started in 1984 but concluded in 2008. The official English translation of the title of this collection clearly shows that these papers describe Chinese maritime history, however, the original Chinese title literally means the history of Chinese ocean development—a topic much broader than maritime history. Thus, a closer examination of the contents of these volumes is required in order to see how the historians relate the concept of 'ocean development' with that of 'maritime activities'. Do they deem these two notions as synonymous? I believe the answer is yes.

Notably, I only introduce the relevant papers in the first three volumes with more details, and then provide a general panorama of the remaining seven volumes. Various papers that are irrelevant to the maritime and ocean affairs have been excluded. For example, the papers written by Shih-chieh Cha (1988) and Mei-rong Lin (1988) are not discussed because they do not mention anything related to maritime and ocean affairs.

The historians mentioned in this chapter did not claim that the ocean was only about maritime trade and migration. Rather, their written works only focus on these uses of the ocean. They represent the ocean as an empty void across which trade and migration can be undertaken. This is the traditional Chinese states' construction of the ocean, and the way of the Indian Ocean Model.

In the first volume of *Zhongguo Haiyang Fazhan Shi Lunwenji*, Chau-nan Chen (2005) wrote its first preface in 1984. In the preface, he identifies the importance of studying Chinese maritime history. He indicates that Chinese academia has focused too much on what happened on the land but little to what occurred on the ocean. He also suggests that the Western countries have benefited from their utilisation of the ocean for transport and colonisation at the expense of other areas of the world. In addition, Taiwan society's members are primarily descendants of immigrants from the Chinese mainland who came across the sea and arrived at Taiwan. He thus argues that it is imperative for Chinese academia to understand how the Chinese people used the ocean in the past.

In addition to the earlier call for the historical study of maritime and migratory affairs, Yih-yuen Li (2005) categorises the historical studies on the Chinese development of the ocean into three main types. The first type studies the role played by the coastal areas of the Chinese mainland in Chinese maritime activities. The second type focuses on the role of offshore islands belonging to China, for example, Taiwan and Hainan Islands. He stresses that the scope of the second type of research should be limited to the maritime activities and migratory processes of these islands. The third type pays attention to the history of trade and migration among the Chinese mainland, offshore islands and foreign countries through sea routes. He also suggests that the third type of studies should be the main focus of the historical research on Chinese ocean development.

Below, I discuss the papers collected in the first three volumes according to Li's (2005) typology. These historians represented and constructed the ocean according to the Indian Ocean Model, given that the ocean was reduced to a space or a surface across or on which migration and trade could be conducted.

Tong-hwa Lee (2005) analyses why the city of Quanzhou became the most pre-eminent harbour for maritime transportation among China, East Asia, South Asia, the Middle East and East Africa during the Song (960–1279) and Yuan (1271–1368) dynasties. Quanzhou is located in the Southeast coastal area of Fujian Province. Geographically, this province became overpopulated beginning in the Song dynasty, and the local Chinese developed a tradition of maritime trade and overseas migration. Nevertheless, the development of Quanzhou during the two dynasties resulted from the convergence of two main maritime sea routes. This convergence highlights the change and development of China's international maritime trade. The largest maritime harbours were located in the eastern coastal areas during the Sui (581–618) and Tang (618–907) dynasties, whereas the largest maritime harbours were found in the southern coastal areas in the Han dynasty (202 BC–AD 220). According to this paper, these two maritime trade routes eventually converged, and the core maritime harbour moved to the Southeast coastal city, Quanzhou.

Yung-ho Tsao (2005) suggests that the maritime policy of Hongwu Emperor (who reigned over China from 1368 to 1398) adopted a proactive attitude towards maritime trade as trade increased the state's revenues. However, having realised that it was extremely difficult to stop smuggling along the Southeast coastal areas, Hongwu Emperor decided to prohibit all maritime trade except with foreign countries granted special imperial privileges. Maritime trade became an instrument of diplomacy. Whether or not Tsao's (2005) explanation for Hongwu Emperor's prohibition of maritime trade is convincing, his focus is not the ocean itself. Furthermore, his analysis of Hongwu Emperor's maritime policy does not indicate that the state was involved in the governance of the sea routes or in the naval protection of maritime trade. The state reigned over by Hongwu Emperor only managed to extract revenues at harbours. The construction of the ocean in his paper thus follows the style of the Indian Ocean Model.

Pin-tsun Chang (2005) analyses the causes of the problem of rampant piracy between 1552 and 1562 in the coastal areas of China in the Ming dynasty (1368–1644). He began his explanation by highlighting that the number of incidents of piracy suddenly increased after 1552. In addition, he clarifies that the people committing piracy were the same as those who undertook smuggling before the problem became uncontrollable in approximately 1552. He indicates that during the rule of the Ming dynasty, maritime merchants often became maritime smugglers because of the prohibition of maritime trade by the Ming state. These merchant-converted smugglers were often joined by many social outcasts. They found their paradise on the sea and earned their living through smuggling with the co-operation of land-based merchants. Because of the illegal nature of maritime smuggling, these smugglers were armed and often organised. However, when the Ming state strengthened its efforts to terminate smuggling, those maritime smugglers were cut off from their terrestrial access to the market for the smuggled goods. Being armed, they simply converted from more peaceful smugglers into violent pirates. In this paper, the ocean is represented as a surface on which smuggling and piracy could occur. Moreover, in his interpretation of the smugglers' view of the ocean, the ocean was also understood as a surface on which their means of living could be obtained. Therefore, this is thus a construction of the ocean corresponding to the Indian Ocean Model.

Tsu-yu Chen (2005) establishes that her paper examines the structure of China's international trade between the 1830s and the 1870s. In addition, this paper examines the changes among China, India and Britain after the 1870s. She attempts to understand the economic importance of China in the world during this period. The paper is included in the collection of the papers on China's ocean development simply because it discusses China's maritime trade.

Hsin-hsiung Chen (1986) begins his analysis by first introducing the development of the technology of shipbuilding in the Song-Yuan era (960–1368).

He then continues to analyse the characteristics of the merchant ships used for long-distance travel in this era. He indicates that the bustling maritime activities, seen in the Song-Yuan era, are a result of the technology of shipbuilding. Once again, the ocean is not discussed, and the focus is on the ships. This is the construction of the ocean in the Indian Ocean Model.

Pin-tsun Chang (1986) discusses the political elites' thoughts about maritime trade and maritime state policy during the Ming dynasty. He notes that the state's maritime policy was lenient before 1368. However, in 1368 the state of the Ming dynasty established the institution of tributary trade and promulgated its policy to prohibit private trade with foreign countries. However, Chinese officials and political elites started to review and reconsider the policy of prohibiting maritime trade. Many of them suggested that the state should stop its prohibition policy because the residents in the coastal areas, especially in the Southeast provinces, depended on maritime trade to supplement their meagre living substances generated from the agricultural sector. He indicates that, for these political elites and the state, the economic benefits of a policy that allowed maritime trade was more for political and security reasons than for fiscal need. This paper does not discuss the ocean itself. The ocean is only rendered as a surface on which maritime trade can be conducted. This is a construction of the ocean belonging to the Indian Ocean Model.

Tong-hwa Lee (1986) identifies that maritime transportation first developed in the Bohai and Yellow Seas to the North and in the East China Sea to the South. He states that the Southeast coastal areas began to develop in the era of the Three Kingdoms (220–280), particularly by the Wu Kingdom. In addition, he notes that the maritime activities increased along with the economic growth in the neighbouring terrestrial areas. Notably, the ocean is once again omitted and constructed in terms of the Indian Ocean Model.

Te-lan Chu (1986) explicates the impact of the Qing state's policy of maritime prohibition, the Great Clearance (*Qian Jie Ling*; Coastal Evacuation Order), implemented between 1661 and 1669. She explicates that the anti-Qing regime in Taiwan, having developed a wide maritime network (and achieved the dominant role in the maritime trade between regimes in Northeast Asia and Southeast Asia), began to suffer from the difficulty of acquiring and selling goods to and from the Chinese mainland after the Qing state launched its campaign of the Great Clearance. Ultimately, the anti-Qing regime could not sustain its military expenses and surrendered to the Qing dynasty (1644–1912). Notably, she evaluates the relations among maritime trade, other economic activities and the finance of the anti-Qing regime in Taiwan. However, the ocean per se is not mentioned and is reduced to a surface on which people can conduct maritime transport and allow trade to be performed. This is the construction of the Indian Ocean Model.

Chung-shen Chang (1986) discusses the pirates in the areas of Fujian and Zhejiang Provinces during the reign of Emperor Jiaqing (who ruled from 1796 to 1820) of the Qing dynasty. The paper examines the relations among the various groups of pirates, their internal organisations, their criminal actions, their terrestrial criminal connections and their financial conditions. The paper is included in the collection of the history of Chinese ocean development, however, it does not mention the ocean itself. It also does not discuss how the pirates utilised the ocean. The ocean is constructed as a surface on which the pirates operated in the vacuum of state power. The ocean is thus constructed in the manner of the Indian Ocean Model in the paper.

Tong-hwa Lee (1988) attempts to examine the politics, economy and international relations of the Nanyue Kingdom (approximately 203 BC–111 BC). He indicates that *Qin Shi Huang* (the first emperor of the Qin dynasty, reigning between 259 BC and 210 BC) conquered the Lingnan area. Consequently, this area became a critical place for the Chinese people (including the Kingdom of Nanyue) to undertake maritime trade. Nevertheless, because this paper only introduces the maritime trade of the Nanyue Kingdom, his focus is on the Kingdom, not the ocean. The ocean is also reduced to a surface across which the people of Nanyue could undertake trade with other terrestrial areas. The paper still constructs the ocean through the lens of the Indian Ocean Model.

Yung-ho Tsao (1988) discusses the relationship between China and Ryukyu during the Ming dynasty. The paper first introduces the system of international relations with China of the countries in East Asia and Southeast Asia during the Ming dynasty. He indicates that the Ming state successfully established a tributary system. In the system, the Ming state favoured Ryukyu. This paper identifies that the Ming state under the Hongwu Emperor required a large quantity of horses for military purposes and Ryukyu produced high quality horses suited for the military use. Therefore, Ming China provided many favours for Ryukyu. This paper also discusses the effects of the Chinese tributary system in the Ming dynasty. However, international trade and the international relations of Ming China were the primary concern of the paper. Maritime trade was just one means for the Ming state to maintain its tributary system. The ocean is thus briefly mentioned, and it is constructed in terms of the Indian Ocean Model.

Tseng-hsin Chang (1988) indicates that the pirates pillaging in the Southeast coastal areas during the Ming dynasty always established their bases in offshore islands. While these pirates' targets of plunder were terrestrial, their bases were always near the island shores. Thus, the pirates indirectly promoted the development of those islands where they were stationed and began to interact with local people. These pirates also provided a living for many marginalised mainland Chinese people. This paper only focuses on the characteristics of the pirates and where they were based. Therefore, the ocean becomes an empty space in which the pirates could traverse and pillage the coastal areas in the Southeast of the Chinese mainland. This construction of the ocean is in the style of the Indian Ocean Model.

Pin-tsun Chang (1988) explicates why maritime trade in the sea areas of East Asia was predominantly controlled by Chinese merchants. An institutional factor was that the Ming state altered its prohibition of maritime trade in 1567. The Chinese merchants were subsequently able to undertake maritime trade through the new port of Zhangzhou in Fujian Province. However, the Europeans were limited to trade with officially designated merchants in China, although they had dominated the maritime trade in other sea areas of the world. For the Europeans, China was closed for trade. They did not enjoy a market in China vast enough to embark on large scale maritime trade on the ocean of East Asia. This paper only analyses the situation and influences of maritime trade across the ocean with China, not the ocean *per se*. Consequently, it is a paper of the Indian Ocean Model.

Te-lan Chu (1986) discusses the situation of Sino-Japanese trade and the development of the port in Nagasaki. She provides details of the goods traded between Chinese and Japanese merchants through Nagasaki. However, the ocean is reduced to a surface across which maritime trade can be conducted. This follows the Indian Ocean Model. The first group of papers, collected in the first three volumes, has been briefly introduced earlier.

The second group, in the first three volumes, is summarised later. Fu-san Huang (2005) examines the influences on Taiwan's society of foreigners undertaking maritime international trade in Taiwan. He indicates three contributions that foreigners made to Taiwan through trade. First, foreigners introduced local Taiwanese produce, such as camphor and tea, into the world market. Second, because local Taiwanese producers gained access to sell more of their goods, various agricultural sectors (tea, for example) expanded. Tea producers in Northern Taiwan prospered accordingly, and the economy in Northern Taiwan gradually surpassed that of Southern Taiwan. Third, foreign companies of international trade disseminated—to the local Chinese people the knowledge and techniques of establishing and managing corporations for international trade. They too started to form their own companies for maritime trade. Once again, this paper can be classified as the Indian Ocean Model.

Ying-chang Chuang and Ling-chih Chou (2005) examine the migratory experiences of the Hakka people into Taiwan. Their paper suggests that those Hakka immigrants successfully established their living patterns similar to their counterparts on the Chinese mainland. The paper also highlights that religious and ancestral worship activities were occasions for the Hakka immigrants to form co-operative agreements to further develop their newly acquired land. The paper constructs the ocean through the Indian Ocean Model. The ocean is not discussed. The authors of this paper merely mention that the Hakka people were immigrants from the Chinese mainland who travelled across the Taiwan Strait.

Yen-hsien Chang's (1986) paper omits the ocean and completely focuses on the Zheng family development process in Hsinchu. The Zhengs began to develop when ZHENG Yongxi became a high-ranking state official. The Zheng family subsequently developed its commercial businesses. Thus, the analysis in this paper constructs the ocean through the Indian Ocean Model.

Yi-chang Liu (1988) indicates that people living in southern China had previously established connections with those living on Taiwan before these areas entered the historical era. People in these two areas had already interacted across the Taiwan Strait. However, the sea is not mentioned. This is a construction of the ocean in the manner of the Indian Ocean Model.

Yen-hsien Chang (1988) argues that the development of Taiwan started in the southern part of the island and gradually spread to the Northern part. However, the specific approach of development usually began from a central area of a region towards the peripheral areas of the rest of the region. For example, the development of Changhua was centred around Lukang; the development of Hsinchu was centred around Tek-kham (Zhuqian). Although he indicates this pattern of Taiwan's development by the Han Chinese immigrants, the ocean that those immigrants had to traverse before they could develop Taiwan remains unmentioned. This paper focuses on what happened on land and, therefore, constructs the ocean in light of the Indian Ocean Model.

Hsueh-chi Hsu (1988) explicates the pattern of people's migration in Penghu. She highlights that while people in Quanzhou and Changzhou moved directly to Taiwan, most of the dwellers of Penghu originated from Kinmen, a small island with similar living conditions, in order to avoid warfare in early Qing under Kangxi Emperor's reign (1654–1722). However, due to natural disasters and the alteration of the living environment, the immigrants in Penghu from Kinmen moved again to Tainan and Kaohsiung in southern Taiwan—where cheap labour was needed. This paper discusses the reasons for people's migration in Penghu, but it does not mention anything about the connection between the migration and the sea. The ocean is not discussed either. Similarly, this is a construction of the ocean in line with the Indian Ocean Model.

Wen-shing Wu (1988) proposes that the migration of mainland Chinese labourers into Taiwan during the Japanese occupation was strictly controlled by the Japanese colonial government. These mainland Chinese labourers were unskilled workers performing difficult jobs with low wages. The number of these labourers in Taiwan also reflected local demands for short-term labour. Although the mainland Chinese workers had to traverse the sea (the Taiwan Strait), this paper does not mention the ocean. Thus, it is also a construction of the Indian Ocean Model.

Fung-te Liao (1988) explicates the problems of piracy, and the perils of the Taiwan Strait, during the Qing dynasty. He clearly states that the transport and communication between the Chinese mainland and Taiwan completely depended on ships. He asserts that the geographical conditions of the sea, and the emergence of piracy, made the sea routes very dangerous, which severely affected the development of Taiwan. During the rule of Jiaqing Emperor, the notorious pirate, CAI Qian, began to pillage in the Southeast coastal and marine areas, which inflicted havoc on the economy of Taiwan—especially during the reign of Daoguang Emperor (ruled from 1820 to 1850). This paper also highlights that these dangerous maritime conditions facilitated the faith of *Ma Zu* (the Goddess of the Sea) in the Southeast coastal regions of the Chinese mainland and Taiwan. The representation of the ocean, presented in this paper, can be categorised as the Indian Ocean Model.

The third group of the papers, in the first three volumes of the collection, is as follows. Lau-fong Mak (2005) discusses the contributions of the Chinese immigrants to the urban development in Singapore and Malaysia. Immigrants are the primary concern, not the ocean (not even how the immigrants moved to Singapore and Malaysia on the sea). The ocean is a void across which migration can be conducted.

Yung-ho Tsao (1986) examines the rise and fall of the status of Baten. He subsequently analyses the role of Chinese merchants in the international trade between Baten and other areas. He suggests that Chinese merchants often handled the transportation and procurement of local special produce. These Chinese merchants also facilitated the immigration of the Chinese in Indonesia. The paper only discusses what people did across the ocean in Baten and how Baten was influenced by international maritime trade during this period. This is a construction of the ocean belonging to the Indian Ocean Model.

Chien-shiung Wu's (1986) paper has a strong humanitarian perspective towards American racism. It highlights at least five types of repercussions on Chinese immigrants in the US caused by the American anti-Chinese movement. First, the US anti-Chinese act promulgated in 1882 prohibiting Chinese immigrants from bringing their wives and children to the US. This made the Chinese immigrant community in the US, a male-dominant society.

Second, the US state prohibited Chinese immigrants from working in many industries that were predominantly advantageous towards the Chinese immigrants, such as mining, railway construction, cigar making, boot making, clothes production, fishing and agricultural production. Consequently, the Chinese immigrants could only earn their living by what would not oppose the interests of the white American people, such as laundry and restaurant services. This subsequently forced the Chinese immigrants to compete against each other. Consequently, Chinese gangsters and organised crime emerged (Wu 1986).

Third, the anti-Chinese movement made most of the Chinese immigrants reside in close proximity. China towns emerged. Fourth, the Chinese immigrants felt that they were excluded and isolated from mainstream American society. Fifth, because the Qing state was too weak to protect their interest, many Chinese immigrants started to support the revolutionary movement led by the Kuomintang on the Chinese mainland (Wu 1986). Notably, the paper does not mention the ocean itself, nor the governance of the ocean by the state. Therefore, it constructs the ocean in the style of the Indian Ocean Model.

Dao-lin Huang (1986) examines the case of the Chinese immigrants in Jamaica (since 1806) to illuminate the history of Chinese migration into the region of the Caribbean Sea. After 1806, the Chinese people began to appear in this area as contract labourers. As they began to manage their own businesses, there was increased Chinese migration into this area. The status of the Chinese immigrants in the area was then elevated. This paper on the history of Chinese ocean development—discussing only the lives of the Chinese immigrants in foreign countries effectively—implies that the ocean is a void which can be crossed for migration. The paper thus constructs the ocean in the way of the Indian Ocean Model.

Chien-shiung Wu (1988) explicates why Chinese labourers went migrated to Cuba. In 1817 and 1835, Spain and Britain reached an agreement banning the trade of slavery. However, Cuba was in severe need of general agricultural labourers and sugar cane plantation labourers in particular. The over-populated China became a source of cheap labour. Chinese labourers were introduced to Cuba as contract workers. Notably, the ocean itself is not mentioned. This perspective of the ocean is limited to migration undertaken across the surface of the ocean. The paper constructs the ocean in line with the Indian Ocean Model.

I have previously introduced each relevant paper on marine affairs contained in the first three volumes of the collection of the history of Chinese ocean development. These papers regard migration and trade, and construct the ocean as does the Indian Ocean Model. Nevertheless, I will provide a panoramic perspective for each of the remaining seven volumes. The panoramic views, and the aforementioned detailed analysis, can sustain the argument that the historians' works have generally constructed the ocean in the style of the Indian Ocean Model.

In his preface to the fourth volume of the collection of papers on the history of Chinese ocean development, Chao-cheng Mai (1991) suggests that China is a terrestrial and maritime-oriented country. Therefore, he argues, it is necessary to examine the history of the following topics for the accumulation and development of the knowledge about the history of the Chinese 'ocean development': maritime trade, international relations and the situations of the international Chinese communities. He also mentions that the history of the development of Taiwan by the Chinese immigrants requires inclusion in the collection of the papers because Taiwan is a successful example of overseas Chinese migratory development. Therefore, it is evident that historians of Taiwan equate ocean development as maritime and migratory activities and growth. The ocean is constructed in the form of the Indian Ocean Model.

In his introduction to the papers in the fourth volume of *Zhongguo Haiyang Fazhan Shi Lunwen Ji*, Chien-shiung Wu (1991) notes that five papers focus on maritime trade, three on the situations of the overseas Chinese immigrants and two on the early development of the Chinese immigrants on Taiwan Island. The last paper discusses the origins of the terms: South China Sea (Nan Hai), East and West Oceans (Dong Yang yu Xi yang), South Ocean (Nan Yang), and Southeast Asia (Dongnan Ya). The papers of this volume can be segmented into three sections: maritime trade, overseas migration and the early development of Taiwan by the Chinese immigrants.

In the first part, the paper on the quantity of silver in China suggests that the loss of silver in China in the first half of the nineteenth century was related to the import of opium into China and a consequence of the reduced international production of silver and gold. The second paper on the states' maritime policies during the Min and Qing dynasties highlights that the concerned societal groups were always more proactive in promoting maritime trade than were the states.

The third paper—using the archives of the Qing dynasty and the British East Indian Company, local historical gazettes and official documents of the Qing state—introduces the transformation in maritime trade through the port of Xiamen between 1727 and 1833. The fourth paper employing the perspective of economic theories of maritime shipping, indicates that the large ships sailing towards Northeast China were from Fujian, the medium ships towards Northern China were from Jiangsu, and the small ships sailing around the Bohai Sea were from Shandong. The fifth paper, examining the trade between China and Africa in the Tang dynasty, identifies the primary types of goods exported and imported through various sea routes and ports.

In the second part, the first paper evaluating the changing types of occupations of the Chinese immigrants in the US—concludes that the Chinese immigrants concentrated on the sectors of laundry, restaurant and grocery services because they were discriminated against by the US anti-Chinese policy. The second paper attempts to understand the Kuomintang sponsored activities among the international Chinese community, and the third paper endeavours to highlight the relations between the overseas Chinese trade company, Taiyi Hao, in Japan and various merchants in Taipei.

In the third part, the first paper provides an overview on the locations of the Hakka settlements, and the second paper elaborates that, under the Japanese colonial regime, mainland Chinese people who had to be hired as contractual labourers in Taiwan were permitted limited access to migrate to this Island.

The fifth volume of the collection of the papers on the history of Chinese 'ocean development' is organised around three themes: external relations through sea routes, maritime trade and overseas Chinese lives.

In the first part, the first paper evaluating the external relations of Jiangsu and Zhejiang areas in the era of the Five Dynasties (907–979) and the second paper examining the situation of Annam under the rule of the Ming dynasty are in effective in dealing with the ocean and human-marine relationships. In the second part, the first paper discusses the foreign relations of the Ming state with the countries surrounding the South China Sea through the institution of tributary trade between 1368 and 1449. The second paper explains why the use of seashells as currency in Yunnan Province collapsed in the seventeenth century and indicates that maritime trade resulted in that collapse.

The third paper investigates the role of Zhapu Harbour in the Sino-Japanese trade during the Qing dynasty. The fourth paper introduces the story of a successful maritime merchant, PAN Youdao, who earned his wealth from the tea trade with foreign countries. The fourth paper examines competitions among the maritime shipping companies in Northern China between 1858 and 1919. Both Chinese and foreign shipping companies were involved in the competitions, and the paper focuses on delineating the advantages and disadvantages that various types of ships could introduce to the competitions.

The fifth paper explains why the copper currency used in Qing China depreciated. This paper highlights that because silver flooded out of China, the value of copper currency vis-à-vis silver dropped. This was a consequence of China being drawn into international maritime trade and the importation of opium. The sixth paper introduces the management of Taiyi Hao and investigates maritime trade between Keelung and Nagasaki. In the third part, the relevant paper deals with the association of Chinese immigrants in New York in the past 100 years. It is a paper on the situations and development Chinese immigrants.

In the sixth volume of the collection of the papers on the history of Chinese ocean development, there are six relevant papers on maritime trade in the Chinese mainland, one on the situations of Chinese immigrants in Taiwan, and one on the overseas Chinese. In the first part, the first paper suggests that there have been three commercial revolutions in China that promoted maritime trade in the coastal areas. The first revolution occurred during the Song dynasty, which gradually shifted commercial activities to the Southeast coastal areas. The second revolution happened in the late Qing period, resulting in the prosperity of the Southeast coast. The third one emerged on Taiwan and Hong Kong in the 1980s and 1990s.

The second paper of the first part explains why Guangzhou gradually lost its critical status in Chinese maritime trade to other harbours in the Southeast coastal areas during the Five Dynasties. The reason was more related to the political situations than the maritime economy. The third paper evaluates the international maritime trade of Guangzhou during the late Song and Early Yuan dynasties. It suggests that the range of Chinese international trade was vast, from Vietnam and the Philippines in the East to Morocco and Spain in the West. Merchants traversed the South China Sea and the Indian Ocean to trade 'across the ocean'.

The fourth paper evaluates the change in the pattern of maritime trade between China and the European countries between 1780 and 1800. It highlights that during these years, the British East India Company gained the predominant status in the maritime trade with China. The fifth paper examines the domestic economic effects of Chinese maritime trade. It shows that the production of tea and silk, the sectors of shipping, railway, insurance and finance began to expand because the Chinese began to embark on maritime trade. The sixth paper deals with the management of the Chinese maritime trade company, Tayi Hao, which was established in Nagasaki, Japan. It details the trade between Taiyi Hao and other companies in Shanghai.

In the second part, there is only one relevant paper. It discusses the relationship between the British East India Company and the political regime of the Zheng family in Taiwan during the early Qing dynasty (1644–1681). The British East India Company dispatched ships to Taiwan in 1670 and the political regime of the Zhengs provided local Taiwan produce in exchange for military supplies. In the third part, the relevant paper elaborates the history of the recruitment of Chinese labourers by Germany to work in Samoa in 1903. This paper indicates that the Chinese immigrants in Samoa were bullied and abused.

Below I will provide an expanded overview on the collection of the papers on the history of Chinese 'ocean development' in the seventh, eighth, ninth and tenth volumes. According to the editor, Shi-yeong Tang (1999), the seventh volume of the collection can be categorised into four types. First, the economy of coastal areas and the history of overseas maritime trade; second, the overseas Chinese; third, the near shore defence and diplomacy; and fourth, the development, maritime shipping and trade of Taiwan.

In the eighth volume of the collection of papers on the history of Chinese ocean development, four categories emerge: first, the history of the development of the coastal areas of the Chinese mainland; second, the history of the Chinese foreign relations; third, the conditions and development of Chinese immigrants in Taiwan; and fourth, the lives of overseas Chinese.

For the first time in these ten volumes of the collection, two papers on the salvage following sea disasters in the Qing dynasty emerged in the eighth volume. These two papers began by the discussion of the salvage and extended it to explicate the foreign relations of the Qing dynasty with Korea and Japan. In those two papers, the Qing state's governance of the sea routes for lifesaving is examined. These two papers construct the ocean in adherence to the Freitian Variation of the Mediterranean Sea Model.

Derived from an academic conference exploring five types of phenomena that happened in the history of the Chinese mainland and Taiwan (first, ship building and navigation technology; second, history of maritime transportation and trade; third, maritime policy and overseas migration; fourth, history of sea disasters; fifth, the history of Taiwan under the Dutch rule and other East Asian history), the ninth volume contains 12 papers and concentrates on first, ships and navigation and second, maritime policy. Notably, the papers on the history of maritime policy do not delve into how previous states governed maritime trade and sea routes. In contrast, they deal with whether the states allowed trade with foreigners at ports and the impact of the policy on the society in the coastal areas. Furthermore, the papers on the states' maritime policy analyse how states managed maritime ships and shipping companies, not sea routes.

The final volume consists of four themes. First, harbours, trade and migration; second, Chinese maritime merchants and overseas trade; third, the maritime trade of Fujian and Taiwan; and fourth, coastal defence and ships.

In the first theme, the first paper discusses Hainan Island and its international trade between the Song and mid-Ming dynasties (960–1577). The focus of the paper is on the ports, merchants and commodities in the commercial circles centred on Hainan Island. The second paper discusses the rise of Zhangzhou in the mid-Ming dynasty (1416–1577) and its emigrants. The third paper examines the situation of Zhangzhou in the late-Ming dynasty (1578–1644) and the merchant ships, maritime trade taxes and coastal defence in Zhangzhou.

Regarding the second theme, the first paper discusses the transformation of the identity of foreign merchants in China in the Tang dynasty. The second paper evaluates the international relations of tributary trade between the Song state, Jiaozhi and Zhancheng. The third paper examines the trade of silk between the Chinese mainland, Taiwan and the Philippines between 1657 and 1686. The fourth paper explores the development of Chinese commercial activities during the era of the Dutch East India Company (1602–1799).

On the third theme, the first paper investigates the opium trade in the coastal areas of Fujian and Guangdong prior to the Opium War (1839–1842). The second paper discusses the sector of prostitution––managed by the Japanese on the Chinese mainland and Taiwan Island—between 1895 and 1945.

On the fourth theme, the first paper provides an analysis of the relationship between the state navy and pirates, and the second paper highlights the historical development of the technology of maritime vessel-building.

In the above introduction to the collection of the papers on the history of Chinese 'ocean development', it is clear that historians in modern Taiwan represented the ocean as an empty space across which migration and trade can be conducted. Their written works constructed the ocean in the manner of the Indian Ocean Model. It is pertinent that in the tenth volume of the collection, published in 2008, the trend of the historians constructing the ocean (in accordance with the Indian Ocean Model) can be found in later, similar academic works. Examples include two newer books respectively authored by Pao-tsum Tai (2011) and Kuo-tung Chen (2013).

The trade and migration, and the use of the ocean, analysed by the historians' works, however, can be classified within the Indian Ocean Model. The historians' works also construct the ocean accordingly. Here, my viewpoint is consistent with Steinberg's (2001) theory of territorial political economy. That is, the use, regulation and representation of a geographical phenomenon (the ocean, in this case) influence each other and shape the construction of this phenomenon.

This perspective also suggests that studying the maritime history of Taiwan cannot produce a distinct Taiwanese oceanic culture that is different from the conventional Chinese maritime experiences. This is because the Han Chinese on Taiwan used and constructed the ocean in the same way as the other Han Chinese in the Southeast coastal areas of the Chinese mainland. Thus, the representations and analyses of the use of the ocean by the Han Chinese on Taiwan provided by historians also inevitably construct the ocean in the conventional Chinese way of constructing the ocean, that is, the way of the Indian Ocean Model.

Historians of modern Taiwan constructed the ocean in the fashion of the Indian Ocean Model. It is now necessary to evaluate how literary works have constructed the ocean.

Literary Works of the Ocean

In this sub-section, I argue that the majority of the literary works about maritime and ocean affairs have constructed the ocean according to the Indian Ocean Model. The literary works will be called oceanic literature from now on in this monograph. Generally, the contemporary oceanic literature on Taiwan adopts an approach that Steinberg (2001: 118) has called 'romantic representation'. I will examine LIAO Hung-chi's (1996) *Taohai Ren* (The Fisherman) as a critical example because it was one of the most influential oceanic novels in the society of modern Taiwan.

When analysing the construction and idealisation of the ocean in Western oceanic literature, within the era of industrial capitalism, approximately between the 1760s and 1960s, Steinberg (2001) offers a binary classification of oceanic literature. The ocean during the era of industrial capitalism was represented and constructed as 'the antithesis of land-space' (Steinberg 2001: 113). This construction of the ocean was affected by classical liberal economic doctrine, which suggests that states should not intervene with trade, but rather invest in the resources that could generate comparative advantages in trade. According to Steinberg (2001), the classical liberal economic doctrine implies that any geographical phenomenon, including the ocean, between national societies conducting trade becomes a great void. The ocean is constructed in the style of the Indian Ocean Model.

The first category of the Western oceanic literature in the era of industrial capitalism, which can be named 'rationalistic literature', perceives 'the ocean as a space resistant to social progress, modernization, and development, and that therefore must be conquered or annihilated (or, short of annihilated, reserved for outcast sailors and non-Western harpoonists)' (Steinberg 2001: 115). The ocean is discursively 'constructed as outside society and the terrestrial places of progress, civilization, and development' (Steinberg 2001: 114).

In contrast, the second category of the Western oceanic literature can be labelled 'romantic literature'. This hails 'the ocean for its wild nature that resisted taming by the forces of modernity' (Steinberg 2001: 118). The ocean is presented as pristine, pure 'space of the sublime' (Steinberg 2001: 118). Influenced by the classical economic doctrine, the romantic oceanic literature understands the land as the basis on which scientific, development-oriented industrial capitalism could flourish. The ocean is perceived as untameable for scientific development and inhospitable to fixed investment. It is worth quoting Steinberg here, the ocean

> represented the points at which rationalism and industrialism–with all their social benefits and attendant evils–stopped, and where Nature–with

> an entirely different and grander set of beauties and evils–began. For romantics, the sea, untainted by the civilization people had built on land, provided an opportunity for individuals and groups to reclaim a more pure humanity.
>
> (2001: 119)

When Steinberg's (2001) binary categorisation is applied, the oceanic literature on Taiwan, reviewed in this sub-section, falls into the second category. However, an exception is Syaman Rapongan's literary works. In his works, a contrast is perceived among the examples of oceanic literature. One of the primary arguments of this chapter is highlighted: how people use, intend to use or urge others to use the ocean can shape their representations and constructions of the ocean. I first explore LIAO Hung-chi's (1996) *Taohai Ren*.

Liao's (1996) novel describes his experiences of working on a fishing boat with the captain, Haiyong Bo. Two points about the models to which his construction of the ocean does not belong can be discussed briefly before I elaborate on the characteristics that enable one to categorise his construction as the Indian Ocean Model.

First, although Liao (1996) describes beautiful oceanic scenes and introduces many interesting marine fishes, the novel does not construct the ocean in the way of the Micronesia Model. Although it appears that the ocean provides fish for fishermen, as described by the novel, the novel provides no narrative that explicates or implies that the ocean is a possessable resource reservoir. In addition, it does not represent the ocean as an entity requiring state sovereignty and governance. In the novel, fishing becomes hunting on the surface of the ocean.

Liao (1996: 205–206; my translation) mentions that 'the sea is a three-dimensional space, big fishes can escape in all directions, and fishing boats have to change direction too in order to chase the big fish'. The fishing boats and fishermen can only chase the fish on the surface of the ocean. Moreover, various chapters of the novel are devoted to the description of the efforts made to harpoon marlins from the fishing boats. The scenes of fish 'hunting' are intensive and nerve-racking in his description. Fish must be found with luck and defeated during the 'hunting battles'. Fish cannot be easily 'retrieved' from the ocean, but must be hunted down with great effort, skill and experience, if the fishermen are lucky enough.

Furthermore, Liao (1996) loves to describe the beautiful marine scenes that one can appreciate on the fishing boats. For example, the whole of page 31 describes this. However, the beauty of the oceanic scenes originates not from the ocean, but from the background of the ocean, such as the stars, mist and sky. The ocean does not become the primary subject of the narrative. Effectively, the ocean per se is the context against which fish 'hunting' could be undertaken.

Second, Liao's (1996) *Taohai Ren* does not construct the ocean in the fashion of the Seldenian Variation. The Seldenian Variation is a sub-type of the

Mediterranean Sea Model. Although this Variation is concerned with fishery, it still requires the governing role of the state, its claim for sovereignty over the nearshore seas and its stewardship of the fisheries. In Liao's (1996) narrative, he suggests that the state did not steward the fishing industry of Taiwan. The marginalised fishermen had to take care of themselves and help each other. This perspective is evident in pages 99, 100 and 101. This construction of the ocean can hardly fit the Seldenian Variation. This is a book describing fishermen's lives on the ocean, not the ocean itself.

Effectively, *Taohai Ren* constructs the ocean as does the Western romantic oceanic literature and in the manner of the Indian Ocean Model. First, throughout the entire book, Liao (1996: 18; my translation) compliments the serenity and tranquillity offered by the ocean:

> floating on this small piece of space on the surface of the ocean, it is a temporary sanctuary. In this space, the complex inter-personal relations are reduced to the simple friendship between me and Haiyong Bo. The rest is the relationships between people and nature, and between people and the ocean. These do not require language or skills; they are serious and straightforward relationships.

The ocean is represented as a pristine, pure surface on which people can temporarily escape from the sophisticated, complicated societies on the land. A man hiding on a boat floating on the sea becomes the antithesis of the collective man-made achievements and development on the land. This is very similar to the Western romantic oceanic literature previously mentioned.

Second, the ocean is depicted as a danger for humans and a place alien to humans. For example,

> the boat is just a very small piece of plate on the ocean, and the fisherman is just one very small spot on this plate. The relationship between the ocean and fishermen shows no equality. Humanity stands no ground and has no basis of confidence. Thus, fishermen very often have to rely on their religion and superstitious practices, and have to use swear words, to enable themselves to follow their own instinct and ease their reluctance caused by the fact that sometimes they have to surrender all of their catch and escape back to the land.
>
> (Liao 1996: 25–26; my translation)

In addition, Liao (1996) emphasises his personal frustration when he fails to capture a large fish. Furthermore, he also uses the whole of the fifth chapter to describe fishing boat disasters. In particular, pages 68 and 69 attempt to show that people do not belong to the ocean and to demonstrate the psychological desire that fishermen had in the events following the breakdown of their fishing boats.

Third, Liao (1996) mentions the supernatural, mystical experiences on the ocean at least four times. The first time is when Haiyong Bo and the author are almost caught in a storm. Haiyong Bo swears at a white ghost before they escape from the sea and return to the land. The second example is when LIAO Hung-chi is trying to catch two big mahi-mahis. His fishhook catches a female mahi-mahi, and the male one refuses to leave. The third time occurs when the author drives the fishing boat alone on the sea, and two marlins attempt to lure the author into the water and kill him in vengeance for various marlins harpooned by other fishermen. The fourth time occurs when the peddle wheel of Haiyong Bo's boat is entangled with fishing nets three times in a single night. Haiyong Bo then sees a black ghost on the sea. Effectively, Liao's (1996) description of these supernatural experiences adds a mystery to his representation of the ocean. Subsequently, the ocean becomes a source of mystical danger that ordinary people should not approach and interfere with.

Fourth, and closely related to the third point, the fishermen are marginalised people in the society of Taiwan, who are strong, courageous and perseverant enough to endure the harsh conditions of fishing on the sea. The sentence that best expresses this notion is found in Haiyong Bo's sarcastic question for the author at the beginning of the author's fisherman life: 'Can't you see the way you are going? You come to this way of fishery!' (Liao 1996: 57; my translation). Effectively, the whole fourth chapter functions to highlight the difficult lives of fishermen.

Liao's (1996) book has constructed the ocean as a surface on which marginalised fishermen risk their lives to 'hunt' fish for living, although the ocean also allows the fishermen to hide and escape from the complex and stressful civilised burden of society on the land. The ocean-space is enchanting, mystical, emancipating and dangerous simultaneously. The fishermen still belong to the land; 'the ocean is ultimately home to the flying fish, as the port is ultimately the end to the fishing boats' (Liao 1996: 18). This is a construction of the ocean similar to the Indian Ocean Model.

I will now proceed to Syaman Rapongan's *Lao Hairen* (The Old Oceanic Man) published in 2009.[3] Syaman Rapongan published his first book, *Leng Hai Qing Shen* (Deep Affection for the Cold Ocean), in 1997, however *Lao Hairen* reveals his more complete and developed ideas about human-marine relations. Through examining *Lao Hairen*, a deeper understanding of the construction of the ocean by his oceanic literature can be gained.

Although he does not directly mention any portion of the ocean as possessable property of a certain tribal family of the Tao people on Lanyu, Syaman Rapongan (2009: 26) does imply that the sea areas surrounding Lanyu, to which the tataras and chinurikurans (the traditional Tao people's handmade wooden fishing boats) can reach, are the customary fishing zones belonging to the whole Tao people.

Second, similar to the Micronesians identified by Steinberg (2001), the ocean provides fish and other forms of living resources for the Tao people. Traditionally, mahi-mahis, flying fish, groupers and many other benthos are the main sources of food and protein. The Tao people, as described by Syaman Rapongan (2009), also trade these resources for other living essentials. In effect, the ocean is a resource reservoir.

Third, in his *Lao Hairen* (and also in his other works), he repeatedly mentions that the *hairenmen* (oceanic men) enjoy the beautiful scenery under the sea. For example, on pages 67, 69 and 70 of *Lao Hairen*, one of the main characters, (Anluomien) often spends time diving and appreciating the scenes under the sea before he shoots fish for food. This vehicle for Anluomien to relax and obtain self-confidence and euphoria. In addition, on page 70, Syaman Rapongan describes the scene that Anluomien is perceiving under the sea. The ocean is represented as a charming place with a fascinating view, not an empty void.

Fourth, fish and other living resources for food, and beautiful scenery for peace of mind, are not the only reasons for oceanic men being so passionate about fishing and diving in the ocean. In Syaman Rapongan's works, he has identified that the culture of the Tao people is fully marine-laden. Every aspect of the Tao culture is closely related to the sea. The religion of the Tao is all about spirits of the ocean. They believe people's souls will be elevated to heaven in the sky where the stars are. Moreover, the Tao culture values the abilities and skills of fish catching and other similarly related techniques. Men are expected to get on and into the sea for fishing, whereas women are left on the land planting taros and attending to the animals. A man is highly regarded if he can catch big, plentiful and valuable fish. Catching ocean fish demands skills, courage, fitness and strength (both physical and mental), as Syaman Rapongan (2009) describes. Therefore, in Tao culture, Tao men are encouraged to be familiar with the ocean since their childhood.

Fifth, Tao men can appreciate the beauty of the ocean and enjoy their stay on the ocean. They prefer to stay on the sea than on the land. The land is necessary for shelter, rest and their family lives, however, the ocean is a place for Tao men to work, to cultivate relationships with family and friends and to gain peace of mind. Furthermore, the oceanic man can be perceived as the extreme form of an ideal Tao man. These *hairenmen* are marginalised people even within Tao society; however, their marginalised status is caused by their lack of ability to earn their living in the monetary economy in the modern industrialised capitalist system. They often resort to alcohol when they are on the land and isolated from others. Their marginal status is not caused by Tao people's traditional culture (Syaman Rapongan 2009).

Nonetheless, they manage to maintain their self-esteem and sustain their lives on, in and by the ocean. 'The ocean always tolerates and provides for them day by day and year by year. Of course, the ocean will not abandon them' (Syaman Rapongan 2009: 21, my translation). The ocean becomes a caring mother, Syaman Rapongan writes, and provided the oceanic men (and other Tao men too) retain innocent hearts, 'the spirits of the ocean will do no harm' (Syaman Rapongan 2009: 132 my translation). The ocean thus 'is like a woman's womb that allows a man to mature with its rhythm' (Syaman

Rapongan 2009: 115; my translation). 'The ocean makes them mature and makes them happy' (Syaman Rapongan 2009: 133; my translation).

Therefore, the ocean is represented and constructed by Syaman Rapongan (2009) according to the Micronesia Model. The ocean, especially the sea areas surrounding Lanyu, is a common, traditional deposit of fish and other living resources belonging, though not exclusively, to the Tao people. Furthermore, the ocean is not merely a resource reservoir, but a place where the Tao men (and oceanic men) can relax, interact, develop and mature. Moreover, the Tao culture encourages Tao men to do so. Their culture and everyday lives are closely interwoven with the ocean. The traditional human-marine relations of the Tao people are primitive, pristine and pure. The relationship reflects the value of harmony between people and the ocean. The ultimate psychological sublimation that a man (and often an oceanic man) can achieve is the mental integration of the man and the ocean.

The oceanic literature of Syaman Rapongan highlights regret over the erosion of the conventional Tao culture. For example, a sarcastic tone is used to describe Tao men using modern boats propelled by diesel engines while thinking themselves clever (Syaman Rapongan 2009). More often, his literature highlights his own normative preferences by contrasting the Tao way with the modernised Han fashion of life. 'Knowledge learned from school is to use rationality to examine the world, whereas the education I received from my father and grandfather is to appreciate the ocean with a harmonious mind' (Syaman Rapongan 2009: 21).

Buttressed by this analysis of the difference, he also distinguishes between the civilised and the savage. For him, 'the people who live in the way closer to the natural environment are called savage (with strong emotions to eco-environment), and the people who live in the fashion farther away from the natural environment are called civilised and would interpret the eco-environment with science' (Syaman Rapongan 2009: 19).

Based on such a distinction, Syaman Rapongan could develop his literary narratives enhancing the ethnic identity and enriching the ethnic discourses of the Tao people. As Hsiao-jung Kuan (1997) identifies, the aboriginal movement on Taiwan, which emerged after 1986, gradually showed its weakness, appeared to lack a goal and forfeited its cultural-ethnic roots. In the early 1990s, Syaman Rapongan, previously a leading figure in the pro-aboriginal movement, returned home on Lanyu. His returning home was an attempt to sustain his own family in the natural economy of his mother tribe. However, his returning home marked his first step to absorb and regenerate traditional Tao culture and to liquidate the stigma entangling the Tao way of life (Kuan 1997).

Syaman Rapongan strives to review, refresh and reformulate the ethnic identity of the Tao people through his anthropological comprehension and literary construction. The concrete results are his novels, and the abstract fruit is a worldview that does not belong to the modernised Taiwan or the contemporary Lanyu struggling to survive. However, this fruit criticises the social and cultural values and practices of the Han Chinese dominating modern Taiwan Island (Hao 1997). Indeed, his oceanic novels highlighting the human-marine relations of the Tao people enriched and enhanced the Tao ethnic identity. The human-marine relations introduced in his works also construct the ocean according to the Micronesia Model.

In this sub-section, I have discussed how the oceanic literature constructed the ocean in modern Taiwan society. LIAO Hongji's novel, as a critical example, constructs the ocean according to the Indian Ocean Model. However, Syaman Rapongan's novels are an exception to the oceanic literature in the society of modern Taiwan. His works construct the ocean according to the Micronesia Model. However, his construction of the ocean was effectively a product of his efforts to reformulate the ethnic identity of the Tao people on Lanyu. The Taiwanese people *qua* Han Chinese people are an object from which he strives to distinguish the Tao. A clear boundary has been made between the 'modernised' and 'industrialised' Han and the 'pristine' and 'natural' Tao. There is little possibility that the Taiwan Independence Movement (TIM) could successfully integrate the construction of the ocean by Syaman Rapongan into the discourses of the TIM.

Conclusion

To conclude this chapter, I have presented an interpretation of many important written works about maritime and ocean affairs of Taiwan through the lens of the Steinbergist Classification. One of the purposes of my interpretation of these written works is to demonstrate the following two theses of the theory of territorial political economy. That is, first, the uses, regulations and representations of the ocean can influence each other and hence affect the constructions of the ocean. Second, how one uses, intends to use, or urges others to use the ocean shapes how one represents and constructs the ocean.

In this chapter, I have explored the written works about maritime and ocean affairs by influential natural scientists, economic and managerial scholars, legal and political intellectuals, historians and literary writers. Remarkably, their representations of the ocean were all shaped by the uses of the ocean of the subjects in their written works or affected by how they urged others to use the ocean.

The natural scientists used and represented the ocean as a reservoir of knowledge. Their studies on the ocean urged the state and other possible economic actors of society to further explore and utilise the ocean. They represented and constructed the ocean in line with the Micronesia Model. The economists and managerial scholars, who paid attention to maritime matters and aimed to maximise profits in maritime-related industries (which used the surface of the ocean), constructed the ocean with characteristics of the Freitian Variation. Those economists and managerial scholars who wrote about fisheries constructed the ocean in terms of the Grotian Variation; and those who wrote about marine resource governance constructed the ocean according to the Micronesia Model.

Absorbing elements of the construction of the ocean by the UNCLOS, legal and political researchers produced analyses about what rights the ROC could have over a certain sea area. Their analyses were about sovereignty disputes over a portion of the ocean. These written works constructed the ocean through the Micronesia Model. Various political scientists of international relations constructed the ocean in an approach like the original Mediterranean Sea Model. They aimed to evaluate how and why a state should use its military force and sea power to dominate and govern a certain sea route of trade and to protect and facilitate military transportation and projection.

Historians on Taiwan generally constructed the ocean according to the Indian Ocean Model. The historians' works typically narrate the maritime trade and migration undertaken across the ocean. The ocean itself is not discussed and is reduced to a surface in those works. The oceanic literature available in modern Taiwan society also constructs the ocean as does the Indian Ocean Model. Syaman Rapongan's works are an exception, whose construction of the ocean belongs to the Micronesia Model and aims to refresh and regenerate the ethnic identity of the Tao people on Lanyu. Aside from his writings, the intellectuals of the humanities described and interpreted the Han people's use of the ocean in a construction pertinent to the Indian Ocean Model. Their constructions of the ocean also fell into this category. In contrast, the interpretation of the Tao people's conventional uses of the ocean facilitated Syanman Rapongan's construction of the ocean in the Micronesian style.

My analysis of the intellectuals' works on maritime and ocean affairs demonstrates that the uses, regulations and representations mutually affect each other and generate corresponding constructions of the ocean. In addition, analysing and understanding the social constructions of the intellectuals-or ideological power holders in Mann's (2012a; 2012b) term—can contribute to the question as to how the society of modern Taiwan constructed the ocean between 1949 and 2016.

Notes

1 This chapter is a modified and expanded version of my journal paper, see Hou (2017).

2 The first edition of *Haiyang Taiwan* is split into four volumes, the second edition into three and the third into three also. Nevertheless, I do not provide a list of the titles of each volume in this chapter. I do not deem it useful for readers to read their titles. They consume too much space and contain little implication for the focus of the current discussion.

3 Although Syaman-Rapongan (2009) does not explicitly define the notion, a *hairen* (oceanic man) is a man who could achieve a harmonious relationship with the ocean and whose life is intertwined with the ocean. The ocean and oceanic man are mentally integrated when he is diving and sailing. Therefore, he can attain tranquility in the ocean.

References

Chang, Chung-shen (1986) 'Qing Jiaqing Nian jian Ming Zhe Haidao Zuzhi Yanjiu' (A Research on the Organisation of Pirates in the Provinces of Fujian and Zhejiang during the Years of Emperor Jiaqing of the Qing Dynasty), in The Editorial Committee for the Collection of Papers of the History of Chinese Ocean Development (ed.), *Zhongguo Haiyang Fazhan Shi Lunwen Ji, di er ji*, Volume 2: Essays in Chinese Maritime History (Taipei; Research Center for Humanities and Social Sciences, Academia Sinica), 161–198.

Cha, Shih-chieh (1988) 'Huangminhua Yundong xia de Taiwan Zhanglao Jiaohu – Yi Nan Bei Jiaohui Xuexiao Shenshe Canbai wei Li' (Taiwan Presbytarian Church uring the Kominka Movement: The Japanese Shrine Worship in the Church Schools in North and South Taiwan as an Example), in Yen-hsien Chang (ed.), *Zhongguo Haiyang Fazhan Shi Lunwen Ji, di san ji*, Volume 3: Essays in Chinese Maritime History (Taipei; Research Center for Humanities and Social Sciences, Academia Sinica), 127–156.

Chang, Kuo-cheng (2013) *Dong Ya Haichuan Lun* (An Essay on the Seapower in East Asia) (New Taipei City; Book Republic).

Chang, Pin-tsun (1988) 'Shiliu zhi Shiba Shiji Huaren zai Dong Ya Shuiyu de Maoyi Youshi' (The Advantages of Trade of the Chinese People in East Asian Waters between the Sixteenth and Eighteenth Centuries), in Yen-hsien Chang (ed.), *Zhongguo Haiyang Fazhan Shi Lunwen Ji, di san ji*, Volume 3: Essays in Chinese Maritime History (Taipei; Research Center for Humanities and Social Sciences, Academia Sinica), 345–368.

Chang, Pin-tsun (2005) 'Shiliu Shiji Zhoushan Qundao de Zousi Maoyi' (The Smuggling qua Trade of Zhoushan Archipelago in the Sixteenth Century), in The Editorial Committee for the Collection of Papers of the History of Chinese Ocean Development (ed.), *Zhongguo Haiyang Fazhan Shi Lunwen Ji, di yi ji, di er ban*, Volume 1: Essays in Chinese Maritime History, 2nd edition (Taipei; Research Center for Humanities and Social Sciences, Academia Sinica), 71–95.

Chang, Tseng-hsin (1988) 'Ming Ji Dongnan Haikou yu Chaowai Fengqi (1567-1644)' (The Pirates in the Southeast in the Era of the Ming Dynasty and their Propensity of Establishing Bases in Overseas Areas), in Yen-hsien Chang (ed.), *Zhongguo Haiyang Fazhan Shi Lunwen Ji*, di san ji, Volume 3: Essays in Chinese Maritime History (Taipei; Research Center for Humanities and Social Sciences, Academia Sinica), 313–343.

Chang, Yen-hsien (1986) 'Taiwan Hsinchu Zhengshi Jiazu de Fazhan Xingtai' (The Pattern of Development of the Zheng Family in Hsinchu, Taiwan), in The Editorial Committee for the Collection of Papers of the History of Chinese Ocean Development (ed.), *Zhongguo Haiyang Fazhan Shi Lunwen Ji, di er ji*, Volume 2: Essays in Chinese Maritime History (Taipei; Research Center for Humanities and Social Sciences, Academia Sinica), 199–217.

Chang, Yen-hsien (1988) 'Hanren Yimin yu Zhongguo Liyu de Tuoken' (The Han Chinese Immigrants and the Development of the Valley of the Zhongguo River), in Yen-hsien Chang (ed.), *Zhongguo Haiyang Fazhan Shi Lunwen Ji, di san ji*, Volume 3: Essays in Chinese Maritime History (Taipei; Research Center for Humanities and Social Sciences, Academia Sinica), 29–60.

Chen, Chau-nan (1986) 'Yingyan' (Foreword), in The Editorial Committee for the Collection of Papers of the History of Chinese Ocean Development (ed.), *Zhongguo Haiyang Fazhan Shi Lunwen Ji, di er ji*, Volume 2: Essays in Chinese Maritime History (Taipei; Research Center for Humanities and Social Sciences, Academia Sinica), no page number available.

Chen, Chau-nan (2005) 'Yinyan' (Foreword), in The Editorial Committee for the Collection of Papers of the History of Chinese Ocean Development (ed.), *Zhongguo Haiyang Fazhan Shi Lunwen Ji, di yi ji, di er ban*, Volume 1: Essays in Chinese Maritime History, 2nd edition (Taipei; Research Center for Humanities and Social Sciences, Academia Sinica), i.

Chen, Jun-yu, Ching-ta Chuang, and Shih-hao Kung (2012) 'Taiwan Guangshangyu Chanye Fazhan zhi Yanjiu' (Development of Ornamental Fish Industry in Taiwan), *Journal of Lan Yan Institute of Technology*, 11, 167–179.

Chen, Kuo-tung (2013) *Dongya Haiyu Yiqian Nian* (One Thousand Years of East Asian Ocean Areas) (Taipei; Yuan-Liou).

Chen, Tsu-yu (2005) 'Yi Zhong Ying Ying Shanjiao Maoyi wei Jizhou Tantao Shijiu Shji Zhongguo de Duiwai Maoyi' (On the International Trade of Nineteenth Century China through Discussing the Triangular Trade between China, India, and Britain), in The Editorial Committee for the Collection of Papers of the History of Chinese Ocean Development (ed.), *Zhongguo Haiyang Fazhan Shi Lunwen Ji, di yi ji, di er ban*, Volume 1: Essays in Chinese Maritime History, 2nd edition (Taipei; Research Center for Humanities and Social Sciences, Academia Sinica), 131–173.

Chiang, Huang-chih (2004) *Guoji Haiyang Fa* (The International Law of the Sea) (Taipei; Sharing).

Chiang, Huang-chih (2013) 'Cong Guoji Haiyang Fa Jianshi "Chungzhiniao" Falü Diwei' (The Legal Status of Okinotoshima), *NTU Law Journal*, 42(3), 432–496.

Chiau, Wen-yan (2001) *Haiyang yu Haian Ziyuan Heli Liyong yu Yongxu Fazhan: 'Zhenghexing Haian Guanli' Lilun yu Shiwu zhi Yanjiu* (Rational Use and Sustainable Development of Marine and Coastal Resources: A Study on Integrated Coastal Zone Management – Theory and Practices) (Taipei; National Scientific Council).

Chiau, Wen-yan (2005) 'Haian Guanli de Taiozhan, Tizhi yu Duice' (Coastal Management: Challenges, Institutions, and Strategies), *RDEC Bimonthly*, 29(4), 33–42.

Chiau, Wen-yan (2007) 'Haiyang Guotu de Guihua: Yuanjing, Fanwei yu Zhuyao Yiti' (Planning for Taiwan's Maritime Territory: Visions, Limits and Major Issues), *City and Planning*, 34(3), 241–272.

Chiau, Wen-yan (2014) 'Jianwang Zhilai: Huigu Kaohsiung yu Taipei Haiyang Xuanyan' (Make an Inference from the Past: A Review of Kaohsiung and Taipei Ocean Declarations), in Wen-yan Chiau (ed.), *Haiyang Taiwan: Yongxu Fazhan: Keyan Jiaoyu yu Haiyang Wenhua* (Oceanic Taiwan: Sustainable Development: Scientific Research and Education, and Marine Culture) (Taipei; National Taiwan Ocean University and Taiwan Research Foundation), 40–67.

Chiu, Forng-chen (2009) 'Haiyang Xinxing Keji zhi Fazhan' (Development of Emerging Oceanographic Technology), in Wen-yan Chiau (ed.), *Haiyang Taiwan: Yongxu Fazhan: Haiyang Chanye yu Keji* (Oceanic Taiwan: Sustainable Development: Marine Industries and Technology Innovation) (Taipei; Taiwan Research Foundation), 236–253.

Chu, Te-lan (1986) 'Qingchu Qianjieling Shi Zhongguo Chuan Haishang Maoyi zhi Yanjiu' (A Research on the Maritime Trade Conducted with the Chinese Ships during the Great Clearance in the Early Qing Dynasty), in The Editorial Committee for the Collection of Papers of the History of Chinese Ocean Development (ed.), *Zhongguo Haiyang Fazhan Shi Lunwen Ji, di er ji*, Volume 2: Essays in Chinese Maritime History (Taipei; Research Center for Humanities and Social Sciences, Academia Sinica), 105–159.

Chuang, Ching-ta (2011) 'Cong Quanqiuhua Siwei Kan Wouguo Yuye Zhengce de Zuoxiang' (The Development of Taiwan Fisheries Policy in Viewing the Concept of Globalization), *Taiwan International Quarterly*, 8(1), 105–124.

Chuang, Ching-ta, Chung-ru Ho, Kwang-ming Liu, Shyh-bing Wang, Wen-yan Chiau, Tien-his Fang, and Min-te Che (2008) *Haiyang Ziyuan Guanli: Lilun yu Shiwu* (Marine Resources Management) (Taipei; Wunan).

Chuang, Ching-ta, Hsiang-his Liu, and Da-chrahn Wu (1998) 'Taiwan Yuanyang Weidiao Yuye Yingying Zeren Yuyezhi zhi Celue Fenxi' (Strategic Analysis to Deep-Sea Tuna Fisheries of Taiwan under Responsible Fishing), *Taiwanese Agricultural Economic Review*, 4(1), 41–71.

Chuang, Ching-ta, Jen-Chyuan Lee, Chian-tshiz You, Hsiang-wen Huang, and Bi Han (2013) *Haiyang Shiwu Gailun* (Introduction of Marine Affairs) (Taipei; Wunan).

Chuang, Ching-ta, Li-te Ho, and Chun-pei Liao (2015) 'Yuchun Zaisheng de Yuanjing yu Shijian: Qiantan Badouzi Shequ de Fudao Jingyan' (Vision to Practice of Fishing Community Regeneration: Experience of Badouzi Community), *Fisheries Extension Report*, 44, 57–84.

Chuang, Ying-chang and Ling-chih Chou (2005) 'Tangshan dao Taiwan: Yige Kejia Zongzu Yiming de Yanjiu' (Tangshan to Taiwan: A Research on the Migration of the Kinship Groups of the Hakka People), in The Editorial Committee for the Collection of Papers of the History of Chinese Ocean Development (ed.), *Zhongguo Haiyang Fazhan Shi Lunwen Ji, di yi ji, di er ban*, Volume 1: Essays in Chinese Maritime History, 2nd edition (Taipei; Research Center for Humanities and Social Sciences, Academia Sinica), 279–333.

Fu, Kuen-chen (2003) *Haiyang Guanli de Falü Wenti* (Ocean Management and Law) (Taipei; Winsoon).

Hao, Yu-hsiang (1997) 'Gudu de Jiushu zhi Di' (A Land of the Salvation for Loneliness) in Syaman Rapongan, *Leng Hai Qing Shen* (Deep Affection for the Cold Ocean) (Taipei; Unitas Publishing), 239–243.

Hou, Kuang-hao (2017) 'Minerva's Dissertations on Neptune: Interpreting the Writings of the Intellectuals of Natural and Social Studies about Maritime and Ocean Affairs in Contemporary Taiwan', *Australian Journal of Maritime and Ocean Affairs*, 9(1), 1–20.

Hsu, Hsueh-chi (1988) 'Penghu de Renkou Qianyi – Yi Baisha Xiang Wuatong Cun wei Li' (The Migration of the Population of Penhu: Wuatong Village of Baisha Twon as an Example), in Yen-hsien Chang (ed.), *Zhongguo Haiyang Fazhan Shi Lunwen Ji, di san ji,* Volume 3: Essays in Chinese Maritime History (Taipei; Research Center for Humanities and Social Sciences, Academia Sinica), 61–93.

Hu, Nien-tsu (1997) *Haiyang Zhengce Zonglun* (A Comprehensive Essay on Marine Policy: Theories and Practices) (Taipei; Wunan).

Hu, Nien-tsu (2002) 'Haiyang Shiwu Bu zhi Sheli: Linian yu Sheji' (The Establishment of Ministry of Marine Affairs: Its Rationale and Design), *National Policy Quarterly*, 1(1), 53–90.

Huang, Dao-lin (1986) 'Jialebi Hai Quyu de Huaren – Qi Yiru ji Shiying de Guocheng yu Tezheng' (The Chinese People in the Area of the Caribbean Sea: the Processes and Characteristics of their Immigrations and Adaptations), in The Editorial Committee for the Collection of Papers of the History of Chinese Ocean Development (ed.), *Zhongguo Haiyang Fazhan Shi Lunwen Ji, di er ji*, Volume 2: Essays in Chinese Maritime History (Taipei; Research Center for Humanities and Social Sciences, Academia Sinica), 327–348.

Huang, Fu-san (2005) 'Qing Ji Taiwan Waishang de Jingying Wenti – yi Meilishi Yanghang wei Li' (The Problems of Management of Foreign Commercial Corporations in Taiwan during the Era of the Qing Dynasty: James Milisch as an Example), in The Editorial Committee for the Collection of Papers of the History of Chinese Ocean

Development (ed.), *Zhongguo Haiyang Fazhan Shi Lunwen Ji, di yi ji, di er ban*, Volume 1: Essays in Chinese Maritime History, 2nd edition (Taipei; Research Center for Humanities and Social Sciences, Academia Sinica), 249–270.

Hwung, Hwung-Hweng (2003) 'Haiyang Diaocha yu Ziliaoku' (The Investigation and Database of the Marine Environment), in Wen-yan Chiau (ed.), *Haiyang yu Taiwan: Haiyang Yongxu Jingying* (Ocean and Taiwan: Sustainable Development of the Ocean), 105–114.

Jan, Man-ser, Chien-yun Cheng, and Ching-ta Chuang (2011) 'Beifang Sandao Haiyang Guojia Gongyuan de Ziyuan Baoyu Fenxi' (Valuation of Environmental Conservation in Three Northern Islets Marine National Park in Taiwan), *Agriculture and Economics*, 47, 1–38.

Jan, Sen, Chih-hao Hsieh, and Chang-feng Dai (2014) 'Haiyang Yangjiu yu Ziliao Zhenghe' [Oceans and Seas Governance Strategy], in Wen-yan Chiau (ed.), *Haiyang Taiwan: Yongxu Fazhan: Keyan Jiaoyu yu Haiyang Wenhua* (Oceanic Taiwan: Sustainable Development: Scientific Research and Education, and Marine Culture) (Taipei; National Taiwan Ocean University and Taiwan Research Foundation), 27–73.

Kuan, Hsiao-jung (1997) 'Cong Shi Nulai dao Syaman Rapongan' (From Shi Nulai to Syaman Rapongan), in Syaman Rapongan, *Leng Hai Qing Shen* (Deep Affection for the Cold Ocean) (Taipei; Unitas Publishing), 5–9.

Lee, Chung-pan, Yang-yih Chen, Hsien-po Tang, Jung-chung Hsu, Shiahn-wern Shyue, Wen-yan Chiau, Chau-chang Wang, and Kuang-che Yang (2003) 'Haian Fanghu Celue yu Yuanze' (Strategies and Principles of Coastal Protection), in Wen-yan Chiau (ed.), *Haiyang yu Taiwan: Haiyang Yongxu Jingying* (Ocean and Taiwan: Sustainable Development of the Ocean) (Taipei; Hushi), 69–82.

Lee, Chung-pan, Ja-jung Kong, and Jon-jer Yang (2009) 'Haian Bainqian yu Haian Zaihai' [Coastal Erosion and Coastal Hazard], in Wen-yan Chiau (ed.), *Haiyang Taiwan: Yongxu Fazhan: Qihou Bianqian yu Haiyang Baohu* (Oceanic Taiwan: Sustainable Development: Climate Change and the Protection of the Ocean) (Taipei; Taiwan Research Foundation), 26–53

Lee, Tong-hwa (1986) 'Haishang Jiaotong yu Gudai Fujian Diqu de Fazhan' (Maritime Transportation and Local Development in Ancient Fujian), in The Editorial Committee for the Collection of Papers of the History of Chinese Ocean Development (ed.), *Zhongguo Haiyang Fazhan Shi Lunwen Ji, di er ji*, Volume 2: Essays in Chinese Maritime History (Taipei; Research Center for Humanities and Social Sciences, Academia Sinica), 59–74.

Lee, Tong-hwa (1988) 'Qin Han Bianju zhong de Nanyue Guo – Lingnan Diqu dui Wai Fazhan Shi Yanjiu zhi Yi' (The Nanyue Kingdom in the Transformation during the Dynasties of Qin and Han: A Research on the Development of the Overseas Relations of the Lingnan Area, Part I), in Yen-hsien Chang (ed.), *Zhongguo Haiyang Fazhan Shi Lunwen Ji, di san ji*, Volume 3: Essays in Chinese Maritime History (Taipei; Research Center for Humanities and Social Sciences, Academia Sinica), 215–242.

Lee, Tong-hwa (2005) 'Song-Yuan Shidai Quanzhou Haiwai Jiaotong de Shengkuang' (The Prosperity of the Overseas Transportation of Quanzhou in the Era of Song-Yuan Dynasties), in The Editorial Committee for the Collection of Papers of the History of Chinese Ocean Development (ed.), *Zhongguo Haiyang Fazhan Shi Lunwen Ji, di yi ji, di er ban*, Volume 1: Essays in Chinese Maritime History, 2nd edition (Taipei; Research Center for Humanities and Social Sciences, Academia Sinica), 1–40.

Li, Yih-yuen (2005) 'Xuyan' (Preface), in The Editorial Committee for the Collection of Papers of the History of Chinese Ocean Development (ed.), *Zhongguo Haiyang Fazhan Shi Lunwen Ji, di yi ji, di er ban*, Volume 2: Essays in Chinese Maritime History,

2nd edition (Taipei; Research Center for Humanities and Social Sciences, Academia Sinica), iii–vii.

Liang, Gin-shuh, Kuang Lin, and Tsung-yu Chou (2003) 'Gouji Shanggang Zhengti Jingying Jixiao Pinggu Moshi: Mohu Jihe Lilun zhi Yingyong' (Integrative Operating Performance Evaluation Model for International Harbour: Application of Fuzzy Set Theory), *Maritime Quarterly*, 12(1), 1–25.

Liao, Fung-te (1988) 'Haidao yu Hainan: Qing Dai Ming Tai Jiaotong Wenti Chutan', in Yen-hsien Chang (ed.), *Zhongguo Haiyang Fazhan Shi Lunwen Ji, di san ji*, Volume 3: Essays in Chinese Maritime History (Taipei; Research Center for Humanities and Social Sciences, Academia Sinica), 191–213.

Liao, Hung-chi (1996) *Taohai Ren* (The Fisherman) (Taichung; Morning Star).

Lin, Kuang and An-sheng Ni (1998) 'Shejie Haiyun Fazhan Qushi yu Wougou Yingyun Huanjing Fenxi' (The Trend of Development in World Shipping Industry and Analysing the Environment of the Management of Shipping Industry in Our Country), *Maritime Quarterly*, 7(4), 1–29.

Lin, Kuang and Chung-hsiung Wang (2000) 'Taiwan Diqu Gangbu Zuoye Shiwu Tantao' (Research of the Practices of the Harbour Operations in the Area of Taiwan), *Maritime Quarterly*, 9(3), 1–7.

Lin, Kuang, Chih-ching Chang, and Shih-liang Chao (2012) *Haiyunxue* (Maritime Shipping), 9th edition (Taipei; Hangmao Wenhua).

Lin, Kuang, Husan-shi Lee, and Yu-jie Wang (2003) 'Yi Caiwu Bilü Pingsu Haiyunye Caiu Yingyun Jixiao zhi Yanjiu' (To Evaluate the Financial Performance for shipping at the Consideration of Financial Ratios by TOPSIS), *Maritime Quarterly*, 7(4), 1–29.

Lin, Kuang, Hsuan-shi Lee, Shuxin Peng, and Yu-jie Wang (1999) 'Qiye Wanlu yu Haiyun Chenglan Yunsongye zhi Yingyong: Yi Tidan Chuli Liucheng wei Li' (The Application of Business Networking on the Industry of Marine Frieght Forwarder: The Procedures of Managing Marine Bill of Lading as an Example), *Maritime Quarterly*, 8(3), 23–37.

Lin, Mei-rong (1988) 'You Jisi Quan dao Xinyang Quan – Taiwan Minjian Shehui de Diyu Guocheng yu Fazhan' (From the Circle of Sacrifice Offering to the Circle of Faith: The Construction and Development of the Geographical Areas in the Society of Taiwan), in Yen-hsien Chang (ed.), *Zhongguo Haiyang Fazhan Shi Lunwen Ji, di san ji*, Volume 3: Essays in Chinese Maritime History (Taipei; Research Center for Humanities and Social Sciences, Academia Sinica), 95–125.

Liu, Yi-chang (1988) 'Shiqian Shidai Taiwan yu Huanan Guanxi Chu Tan' (A Preliminary Examination of the Relationship between Taiwan and Southern China in Pre-historical Era), in Yen-hsien Chang (ed.), *Zhongguo Haiyang Fazhan Shi Lunwen Ji, di san ji*, Volume 3: Essays in Chinese Maritime History (Taipei; Research Center for Humanities and Social Sciences, Academia Sinica), 1–27.

Mai, Chao-cheng (1991) 'Xu' (Preface), in Chien-shiung Wu (ed.) *Zhongguo Haiyang Fazhan Shi Lunwen Ji, di si ji*, Volume 4: Essays in Chinese Maritime History (Taipei; Research Center for Humanities and Social Sciences, Academia Sinica), i–ii.

Mak, Lau-fong (2005) 'Zaoqi Huaren Shehui Zuzhi yu Xin Ma Chengzhen Fazhan de Moshi' (Early Chinese Social Organisations and the Pattern of Urban Development in Singapore and Malaysia), in The Editorial Committee for the Collection of Papers of the History of Chinese Ocean Development (ed.), *Zhongguo Haiyang Fazhan Shi Lunwen Ji, di yi ji, di er ban*, Volume 1: Essays in Chinese Maritime History, 2nd edition (Taipei; Research Center for Humanities and Social Sciences, Academia Sinica), 367–404.

Mann, Michael (2012a) *The Sources of Social Power*, Volume 1: A History of Power from the Beginning to AD 1760, 2nd edition (Cambridge; Cambridge University Press).

Mann, Michael (2012b) *The Sources of Social Power*, Volume 2: The Rise of Classes and Nation-states, 1760–1914, new edition (Cambridge, Cambridge University Press).

Shao, Kwan-tsao (2003) 'Haiyang Shengtai Baoyu' (The Protection of Marine Eco-system), in Wen-yan Chiau (ed.), *Haiyang yu Taiwan: Haiyang Yongxu Jingying* (Ocean and Taiwan: Sustainable Development of the Ocean) (Taipei; Hushi), 87–113.

Shao, Kwang-tsao (2009) 'Haiyang Gongneng zhong Youguan Haiyang Baohuqu de Haushe Wenti' (Questions Related to the Establishment of Marine Protected Areas), in Wen-yan Chiau (ed.), *Haiyang Taiwan: Yongxu Fazhan: Haiyang Chanye yu Keji Chuangxin* (Oceanic Taiwan: Sustainable Development: Marine Industries and Technology Innovation) (Taipei; Taiwan Research Foundation), 199–205.

Shao, Kwang-tsao (2014) 'Haiyang Baohuqu yu Shengtaixi de Yuye Guanli' (Marine Protected Areas and Eco-system Approach to Fisheries), in Wen-yan Chiau (ed.), *Haiyang Taiwan: Yongxu Fazhan: Lance Chanye yu Huanjing Baoyu* (Oceanic Taiwan: Sustainable Development: Blue Industries and the Protection of Environment) (Taipei; National Taiwan Ocean University and Taiwan Research Foundation), 105–114.

Song, Yann-huei (2015) 'Legal Status of Taiping Island under the United Nation Convention on the Law of the Sea', *Korean Journal of International and Comparative Law*, 3, 115–138.

Steinberg, Philip E. (2001) *The Social Construction of the Ocean* (Cambridge; Cambridge University Press).

Syaman Rapongan (1997) *Leng Hai Qing Shen* (Deep Affection for the Cold Ocean) (Taipei; Unitas Publishing).

Syaman Rapongan (2009) *Lao Hairen* (The Old Oceanic Man) (Taipei; Ink Publishing).

Tai, Pao-tsum (2011) *Taiwan de Haiyang Lishi Wenhua* (Oceanic Culture and History of Taiwan) (Taipei; Yushan She).

Tang, Shi-yeong (1999) 'Bianzhe Xu' (Editor's Preface), in Shi-yeong Tang (ed.), *Zhongguo Haiyang Fazhan Shi Lunwen Ji, di qi ji*, Volume 7: Essays in Chinese Maritime History (Taipei; Research Center for Humanities and Social Sciences, Academia Sinica), iii–iv.

Tsao, Yung-ho (1986) 'Ming Muo Huaren zai Zhuawa Wandan de Huodong' (The Chinese Activities in Baten, Java in the Era of Late Ming Dynasty), in The Editorial Committee for the Collection of Papers of the History of Chinese Ocean Development (ed.), *Zhongguo Haiyang Fazhan Shi Lunwen Ji, di er ji*, Volume 2: Essays in Chinese Maritime History (Taipei; Research Center for Humanities and Social Sciences, Academia Sinica), 219–247.

Tsao, Yung-ho (1988) 'Ming Hongwu Chao de Zhong Liu Guanxi' (The Sino-Ryukyu Relations under the Rule of the Hongwu Emperor in the Ming Dynasty), in Yen-hsien Chang (ed.), *Zhongguo Haiyang Fazhan Shi Lunwen Ji, di san ji*, Volume 2: Essays in Chinese Maritime History (Taipei; Research Center for Humanities and Social Sciences, Academia Sinica), 283–312.

Tsao, Yung-ho (2005) 'Shi Lun Ming Taizu de Haiyang Jiaotong Zhengce' (An Essay on the Maritime Policy of Hongwu Emperor), in The Editorial Committee for the Collection of Papers of the History of Chinese Ocean Development (ed.), *Zhongguo Haiyang Fazhan Shi Lunwen Ji, di yi ji, di er ban*, Volume 2: Essays in Chinese Maritime History, 2nd edition (Taipei; Research Center for Humanities and Social Sciences, Academia Sinica), 41–70.

Wu, Chien-shiung (1986) 'Meiguo Paihua Yundong ji Qi dui Zhongguo Yimin de Yingxiang' (The Anti-Chinese Movements in America and Its Repercussions on the

Chinese Immigrants), in The Editorial Committee for the Collection of Papers of the History of Chinese Ocean Development (ed.), *Zhongguo Haiyang Fazhan Shi Lunwen Ji, di er ji*, Volume 2: Essays in Chinese Maritime History (Taipei; Research Center for Humanities and Social Sciences, Academia Sinica), 289–325.

Wu, Chien-shiung (1988) 'Shijiu Shiji Qianwang Guba de Huagong (1847–1874)' (The Chinese Labourers Emigrating towards Cuba in the Nineteenth Century (1847–1874)), in Yen-hsien Chang (ed.), *Zhongguo Haiyang Fazhan Shi Lunwen Ji, di san ji*, Volume 3: Essays in Chinese Maritime History (Taipei; Research Center for Humanities and Social Sciences, Academia Sinica), 417–476.

Wu, Chien-shiung (1991) 'Bianzhe Xuyan' (Editor's Foreword), in Chien-shiung Wu (ed.), *Zhongguo Haiyang Fazhan Shi Lunwen Ji, di si ji*, Volume 4: Essays in Chinese Maritime History (Taipei; Research Center for Humanities and Social Sciences, Academia Sinica), iii–vii.

Wu, Wen-shing (1988) 'Ri Ju Shiqi Lai Tai Hua Gong zhi Tantao' (An Examination of the Immigrant Labourers from Mainland China in Taiwan), in Yen-hsien Chang (ed.), *Zhongguo Haiyang Fazhan Shi Lunwen Ji, di san ji*, Volume 3: Essays in Chinese Maritime History (Taipei; Research Center for Humanities and Social Sciences, Academia Sinica), 157–189.

Yu, Kien-hong (2015) *Ocean Governance, Regimes, and the South China Sea Issues* (Singapore; Springer Science+Business Media Singapore).

4 The Social Constructions of the Ocean by the Military Sources of Power in Modern Taiwan[1]

Introduction

In the previous two chapters, I have employed the Steinbergist Classification and Steinberg's (2001) theory of territorial political economy to explore the social constructions of the ocean formed by the state and ideological sources of power in modern Taiwan society. Through this exploration, one of the theses in the theory of territorial political economy has been demonstrated: how one uses, intends to use or urges others to use the ocean affects one's construction of the ocean. In turn, the use, regulation and representation of the ocean mutually influence each other and shape the construction of the ocean.

This chapter examines the social constructions of the ocean formed by the military sources of power on modern Taiwan. While Chapter 3 deals with a phenomenon that has plenty of examples, mainly after the 1980s, my discussion here traces the history back to the era of the 1950s. Therefore, this chapter needs to present more historical experiences and, thus, exemplifies another thesis of the theory of territorial political economy: the social construction of the ocean is also influenced by the geographical conditions of the ocean and the material organisation of society.

In this chapter, I suggest that the Taiwan Strait provides a geographic environment for the defence of Taiwan Island against the People's Liberation Army (PLA). However, the state, the Republic of China (ROC), stipulated diverse grand strategies in various phases after 1949. Correspondingly, the military forces under the ROC's command used, regulated and represented the ocean according to different strategies. Moreover, these diverse constructions of the ocean by the military forces not only corresponded to the grand strategies established by the state but were also closely related to the military capabilities of the ROC and the Chinese Communist Party (CCP) for utilising the ocean, in particular naval power. Military and naval weapons possessed by the ROC's and CCP's military forces changed from phase to phase; thus, how ROC's military constructed the ocean also varied.

Above all, the Taiwan Strait was an ongoing geographical factor influencing the state's grand strategies and the armed forces' military strategies in

DOI: 10.4324/9781003291343-4

the phases discussed in this chapter. Furthermore, the grand strategies of the ROC along with the military technologies and weapons of the two sides on the Taiwan Strait influenced the material organisation of society. Affected by the above factors, how the military socially constructed the ocean during the different phases varied accordingly.

Notably, Steinberg (2001) suggests that the naval forces of advanced industrialised capitalist countries constructed the ocean as they did in the European mercantilist era (roughly between the fifteenth and eighteenth centuries) in the style of the original Mediterranean Sea Model. He observes that the era of industrial capitalism began approximately in 1760 and lasted for around two hundred years and that the navies in the industrial capitalist era did not follow the overall construction of the ocean by other marine user groups—the Indian Ocean Model. He argues that such an overall construction was a result of the concurrent material organisation of society, industrial capitalism, which required developing terrestrial areas for capital investment and reproduction. He also claims that the naval forces in the industrial capitalist era resisted the overall construction of the ocean and employed Alfred Thayer Mahan's (2007 [1890]) theory of sea power to justify their requests for a higher defence budget and to maintain their preeminent status in the three services of the armed forces.

According to Steinberg (2001), navies have played no effective role in securing the command of the sea since the industrial capitalist era. He maintains that since the Napoleonic Wars (1793–1815), European and American naval fleets have not actively pursued pitched sea battles against each other. Rather, they have mainly been involved in interrupting and protecting maritime transportation for civil and military purposes. Examples include Napoleon's grand strategy of closing the continental market for Britain, German military strategy of submarine warfare and American military strategy of leapfrogging during the Second World War (1939–1945). Thus, in Steinberg's (2001) view, since the era of industrial capitalism, decisive sea battles have been avoided and the status of naval forces has been downgraded and converted into that of saboteurs and bodyguards of maritime transport.

I consider that the above view is simplistic. First, examples employed by Steinberg (2001) do not prove the redundancy of naval fleets and the lack of necessity of building naval power in the context of industrial capitalism. On the contrary, states with dominant and powerful naval fleets were such that weaker opponents strove to avoid being crushed decisively in naval battles. The states controlling weaker navies naturally and rationally pursued military strategies to undermine the socioeconomic infrastructure of the mighty fleets of their opponents, such as submarine warfare. Paradoxically, possessing overwhelming naval power could result in the difficulty, if not impossibility, of destroying the weaker enemy navy that endeavoured to avoid a sea battle. Nevertheless, this difficulty did not prove building strong naval fleets to be unnecessary.

Second, the fact that the weaker navy rationally avoided an all-out clash against the dominant naval enemy was because the stronger navy was too powerful for its weaker opponent to launch any credible and direct military assault against it; that is, the reason for the avoidance of pitched naval battles was that one side enjoyed overwhelming naval power. For example, had the British navy not been so powerful, it would not have been imperative for the German counterpart to implement so rigorously the strategy of submarine warfare in the First World War (1914–1918). The prevention and absence of pitched sea battles could be a result of the very existence of a navy that had already secured the command of the sea, or at least dominant status at sea. It could still be a rational choice for a state with a strong economic base to build dominant or even overwhelming naval fleets if the state desired a relatively less risky and safer maritime environment.

Third, a weaker navy could evade sea battles and adopt a naval strategy of 'fleet in being'. If used properly, this naval *qua* military strategy constituted an effective approach to sea denial for defence (Corbett 1911). Submarine warfare was not the only choice for weaker naval forces. Stationed in well-protected harbours or staying in strategic sea areas in an evasive manner could also hinder the movement of the stronger naval forces. A weaker 'fleet in being' poising to attack the weakness of its stronger naval opponent could affect the calculation of military tactics of the latter and tether the enemy to a certain sea area, in which the stronger fleet had to constantly guard against the weaker. The German-British naval stalemate during the First and Second World Wars discussed by Steinberg (2001), to a large extent, were a result of such a strategy. Notably, this point is not considered by Steinberg (2001). Modern navies in the industrial capitalist era did not necessarily need to adhere to the Mahanian doctrine to justify their existence. Neither did the absence of decisive sea battles mean that naval forces would inevitably be reduced to escorting maritime transport.

Fourth, historical experiences contradict Steinberg's (2001) generalisation about the absence and unimportance of decisive sea battles in the era of industrial capitalism. I have not found his interpretation of the Battles of the Nile and Trafalgar nor his explication of the influence of the Midway Island Battle. Could the United States (US) Navy have adopted the naval strategy of leapfrogging had it lost the Midway Island Battle? Critical, decisive sea battles did not frequently occur in the history of war. However, does this mean that navies can be deemed redundant and reduced to a secondary role? I suggest not.

Above all, having suggested several points that can polemicise Steinberg's (2001) claim that building a strong navy in the modern era was unnecessary, my analysis in this chapter still reinforces his account. He is correct that the naval construction of the ocean in the industrial capitalist era was in line with the original Mediterranean Sea Model. He is also correct to claim that the economic sources of power during that period generally constructed the ocean as the Indian Ocean Model. However, the general naval construction of the ocean was by no means the result of naval resistance against change and refusal to accept the minor role of military escorts. Two critical factors influencing how navies constructed the ocean emerge: first, the grand strategies of states controlling the navies and second, the changing balance of military power between the opposing navies.

This chapter aims to shed light on this viewpoint with an historical analysis. The ROC had diverse grand strategies in different phases and commanded its military sources of power accordingly between 1949 and 2016. Furthermore, military technologies and weapons possessed by the armed forces of the ROC and the People's Public of China (PRC), respectively, were improved and upgraded at different paces. The balance of military power between these two political entities was changing. Consequently, the aforementioned two factors—the grand strategy and the balance of military power across the Taiwan Strait—influenced how the ROC's military forces used, regulated and represented the ocean variously under different conditions.

The layout of this chapter is as follows. The next section describes the military clashes across the Taiwan Strait between 1949 and 1971. The PRC was admitted by the United Nations (UN) in 1971. This description explicates the grand strategy of the state and the corresponding constructions of the ocean by the military forces. During this period, JIANG Jieshi (also known as CHIANG Kai-shek; hereafter JJS) did not abandon the strategy of retaking the Chinese mainland by force. The ROC's navy was stronger than the PLA Navy (PLAN) and was reinforced by the US Navy. I, therefore, suggest that the construction of the ocean by the military sources of power was one of the original Mediterranean Sea Model, although it was changing towards the Indian Ocean Model after 1965.

The third section explicates the grand strategy of the state and the construction of the ocean by the military between 1971 and 1991. The state's grand strategy during this period was staunchly defensive. The ROC adopted a militarily defensive but ideologically offensive grand strategy. More specifically, the state did not renounce its attempt to reclaim the Chinese mainland but shifted its military focus to the defence of Taiwan, Kinmen and Matsu. The ROC's military power constructed the ocean in the manner of the Indian Ocean Model.

The fourth section elaborates the state's grand strategy between 1991 and 2016, which became a purely defensive one that had renounced any intention to reclaim the Chinese mainland and remove the CCP regime. Moreover, after 1991, it became obvious that the PLA had military superiority over the armed forces of the ROC. Between 1991 and 2000, the ROC's military forces gradually altered their construction of the ocean to the original Mediterranean Sea Model. The ROC's grand strategy was formally converted into deterrence-centred defence (emphasising deterrence more than protection) between 2000 and 2008. This change did not transform the military's construction of the ocean, which remained as the original Mediterranean Sea Model. After 2008, the military under the rule of the Kuomintang (KMT) diluted the hostile connotation of deterrence and changed its construction of the ocean into one of the Indian Ocean Model up to 2016. This section provides the historical account necessary for comprehending the development of those different grand strategies and, hence, the construction of the ocean by the military.

Above all, the state's grand strategies and the changing balance of military power across the Taiwan Strait in various phases could precipitate a corresponding social construction of the ocean by the armed forces. The fifth, concluding section is used to clarify the above point with comparisons of the situations in the three phases.

Nevertheless, a clarification of what strategies, doctrines and tactics mean in this monograph can facilitate a clearer explication of the theme in this chapter presented later. Generally, in geopolitical theories, there is a hierarchy amongst strategies, doctrines and tactics. At the top is the grand strategy, followed by military strategy, military doctrine, which is one rank further below, and at the bottom, military tactics.

Such a conceptual hierarchy and its components are ideational instruments that help researchers to organise their observation and data. In reality, they may not be the products of deliberate and careful design. They may simply be a series of intentions, actions and consequences of the actions that can be categorised into grand strategies, military strategies, doctrines and tactics through these conceptual lenses.

As Stephen Biddle (2007: 461) identifies, grand strategy can be defined as 'the state's ultimate security objectives', and he points out methods to achieve the objectives via military, economic, ideological and geopolitical approaches. An example is the US attempting to contain the expansion of the Soviet Union during the Cold War (1947–1991).

Military strategy, also known as theatre strategy, is regarding the direction of actions of the largest military formations, such as army groups, naval fleets or aircraft groups of wings (Biddle 2007). It prescribes how armed forces accomplish their goals established by the grand strategy in a certain theatre of war, for example, Mahan's (2007) emphasis on building powerful battle fleets and controlling commercial sea lanes after decisive naval battles.

Military doctrine, also known as the operational art, provides a prescription for military forces that attempt to achieve theatre strategic objectives by connecting a set of battles, engagements or strikes (Biddle 2007). A military doctrine 'concerns the activities of large formations (such as army divisions or corps, naval task forces, or wings of aircrafts)' (Biddle 2007: 462). For example, during Second World War, the US army doctrine considered offensive campaigns as a series of sequential battles that occurred in four stages: concentration, breakthrough, exploitation and pursuit (Biddle 2007).

Finally, military tactics prescribe 'how the individual battles or airstrikes that make up a campaign are to be conducted, and ordinarily dictate how small units are employed (e.g., platoons, companies, battalions, ships, or squadrons)' (Biddle 2007: 463). For example, before First World War, European armies employed infantries as the primary arm while artillery branches played a supportive role in battles (Biddle 2007).

Grand Strategy and the Military's Construction of the Ocean between 1949 and 1971

The ROC's Grand Strategy between 1949 and 1971

Following the definition of the grand strategy provided earlier, I introduce the four aspects of the ROC's grand strategy between 1949 and 1971. Regarding the ideological aspect of the ROC's grand strategy, JJS defined the Chinese Nationalist-Communist Civil War (1946–1949) as a zero-sum struggle between *San Min Zhuyi* (the Three Principles of the People; hereafter TPP) formulated by SUN Yixian (also known as SUN Yat-sen, the founder of the KMT and the ROC) and Marxism-Leninism-Mao Zedong Thought (hereafter, Chinese Communism) embraced by the CCP (Chung 2012). The TPP include the ideals of Chinese nationalism and self-determination, liberal constitutional democracy and state-governed capitalist economy.

Although JJS imposed the Temporary Provisions Effective during the Period of National Mobilization for Suppression of the Communist Rebellion (hereafter Temporary Provisions) in 1948 and established an authoritarian dictatorship after the ROC retreated to Taiwan, he always claimed the superiority of the TPP for building an independent, democratic and prosperous China. Moreover, because of the ideological incompatibility between the TPP and Chinese Communism, JJS, as an ardent supporter of the former, uncompromisingly devoted himself to the eradication of communism from the Chinese mainland (Chung 2012).

With regard to the economic aspect of the ROC's grand strategy, JJS understood the urgency of improving Taiwan's economic conditions, which were considered the basis of the military strategy to reclaim the Chinese mainland. Although JJS maintained a military force of 600,000 soldiers and spent approximately 70 per cent of the state budget on the military in the early 1950s (Lin 2016: 275; Wade 2004: 265), he was still able to implement economic, industrial policies to stabilise Taiwan's economy.

Although these policies were designed to establish a strong logistic base for military counterattacks against the CCP on the Chinese mainland, they became the initial steps for later economic growth and industrial development. To forestall infiltration by the Chinese Communists, JJS authorised his lieutenant, CHEN Cheng, to begin land and currency reforms in April and June of 1949, respectively. A special committee for managing industrial production in all state, provincial and private enterprises was also set up (Chang 2001: 521).

Moreover, JJS ordered the transport of gold worth US$1 billion to Taiwan between December 1948 and May 1949, which was the basis for issuing a new currency, the New Taiwan dollar, in June 1949 and reducing inflation (Chang 2001: 530). Additionally, having accepted US financial aid after 1951, the ROC succeeded in carrying out land reform. The ratio of self-sufficient land-owning farmers to all agricultural workers increased from 36 per cent in 1949 to 83 per cent in 1957 (Chang 2001: 531).

Furthermore, between 1952 and 1966, the percentage of the population dedicated to agriculture decreased from 55 per cent to 43.4 per cent; that of the industrial sector increased from 17 per cent to 23.4 per cent; and that of the service sector grew from 28 per cent to 33.2 per cent. On average, the annual economic growth rate between 1953 and 1981 was 8.9 per cent (Chang 2001: 533). The economy continued to develop after 1971, but the foundation had been established earlier as a part of the ROC's grand strategy against the Chinese Communists.

With respect to the geopolitical aspects of the ROC's grand strategy, domestic and international components were of equal importance. Domestically, JJS began to reform the organisation of the KMT. Critical positions were rearranged and assigned to his trusted subordinates. After the Seventh National Congress of the KMT in October 1952, JJS secured his domination over the KMT and largely increased its organisational discipline (Chang 2001: 522–523). In addition, JJS imposed martial law on Taiwan in May 1949, enhancing the power of the Taiwan Garrison Command. The task of removing the threats of communist infiltration and the Taiwan Independence Movement (TIM) was assigned to this military organisation. In a nutshell, JJS strove to secure the home front soon after the ROC withdrew to Taiwan.

Internationally, although the US reversed its ROC policy before the Korean War, the US eventually announced the policy of neutralising the Taiwan Strait and dispatched its Seventh Fleet to protect the islands of Taiwan on 25 June 1950 (Chang 2001: 526). With fewer than 30,000 cannons, approximately 60 aeroplanes and 130 naval vessels, the PRC did not possess sufficient military power to defeat the US fleet at that time.

After the PRC entered the Korean War on 25 December 1950, the US began to provide financial aid and weapons to the ROC's military forces (Chang 2001: 554–555). This US protection, however, created an awkward situation for the ROC. On the one hand, it prevented the PRC from invading Taiwan Island; but on the other, it prohibited the KMT from launching a counterattack to reclaim the Chinese mainland (Chang 2001). JJS attempted to resolve this situation with a bold strategic move of deploying more than 150,000 soldiers on Kinmen and Matsu islands. Two important effects were generated by this deployment. First, it effectively forced the US to extend its naval protection to Kinmen and Matsu, even though US military support was not official. This was because the fall of these two islands would have meant a fatal blow to the remaining military forces in Taiwan. Second, having a sizeable army force stationed on Kinmen and Matsu posed a substantial threat to the southeast coastal areas of the Chinese mainland. The islands effectively provided access for the ROC's military forces to launch a counterattack against the PLA (Chung 2012).

The Construction of the Ocean by the ROC's Military Forces between 1949 and 1971

To understand the military's construction of the ocean between 1949 and 1971, in addition to the ROC's grand strategy, how the military used (and claimed to use) the ocean to obtain the goals of the grand strategy during this period is also relevant. During the Second World War, while defending against the Japanese amphibious assault and preventing the Japanese navy from controlling the Yangzi River, the ROC Navy (ROCN) lost nearly all its vessels. Many were destroyed by Japan, while others were sunk in the Yangzi River by the ROCN itself to block the advance of Japanese naval ships.

The ROCN suffered terrible losses in halting Japan's advancement along the Yangzi River. The ROC had decided to reinstate and reorganise its navy during wartime, dispatching many junior naval officers to the US and the United Kingdom (UK) to learn the latest naval *cum* military strategies and technologies. The ROCN also accepted four gift ships (one from the US and three from the UK) in 1941 (Liu 1994: 44). After the war, the ROC took over the surrendered Japanese vessels and rented, borrowed and received ships and boats from the US and UK. The ROC conducted the reinstatement and reorganisation through three steps taken in June 1947, April 1948 and August 1949. The ROCN then possessed around 240 vessels and assigned them to four fleets based in Shanghai, Qingdao, Taiwan and Hainan (Liu 1994: 47).

During the Nationalist-Communist Civil War (1946–1949), the ROCN focused on the bombardment of the shores, the interception of enemy maritime movements and the maritime transportation for military supplies. Towards the end of the civil war, the ROCN concentrated its capacity for transporting military and civilian personnel and movable materials and resources to Taiwan Island (Chang 2001; Chung 2003a).

For example, Admiral LI Yuxi (2013) recalls that, when he was the captain of the Taikang destroyer in 1948, he accomplished many important missions, including cooperating with the 26th Division of the ROC Army (ROCA) to reclaim Xuejia Island, intercepting and destroying a Chinese Communist ship on the shores of Moye Island, bombarding the communist naval base and ships at Rongcheng Bay, supporting the defence of Qinhuang Island, intercepting the communist reinforcement through sea lanes, shielding the withdrawal of the 52nd Army of the ROCA, and covering the retreat of the ROCA from Hulu Island.

Notably, as Li (2013) mentions, since late October 1945, the Chinese Communists had begun utilising sea lanes between the Shandong and Liaoning provinces for the transportation of military personnel and materials. However, because of factional struggles in the KMT, the ROCN under the command of CHEN Shaokuan did not complete its assigned mission of intercepting the CCP's maritime transport. The ROCN had made little, if any, effort to stem the flow of military personnel and materials until the task was assigned to LI Yuxi in August 1946.

The implications of Li's (2013) explication of the above situation are two-fold. First, if the ROCN had begun prohibiting the communists from using the sea lanes earlier, the fighting capacity of the PLA could have been lower. Second, although the ROCN did not prevent the CCP from using the sea, the ROCN certainly understood the importance of the military domination of the ocean and sea lanes, and, in turn, constructed the ocean in the manner of the original Mediterranean Sea Model.

Having withdrawn to Taiwan in 1949, JJS first followed the advice of the retired US four-star admiral, Charles M. Cooke, Jr. and concentrated the ROC's military power on protecting Taiwan in the early 1950s (Lin 2016). JJS himself and other high-ranking Nationalist military officers expected the communist invasion of Taiwan to take place between July and September 1950 (Chung 2012: 77). The outbreak of the Korean War allowed the ROC to enjoy the protection of the US Seventh Fleet, giving JJS the time and space to implement his grand offensive strategy of reclaiming the Chinese mainland (Chung 2012). However, during the Korean War, he privately oversaw the tasks of reorganising the KMT, reequipping and readjusting the Nationalist armed forces and consolidating his political position as his first priorities (Lin 2016; Yang 2003a).

Despite these priorities in the early 1950s, JJS did not deviate from his grand strategy of recovering the Chinese mainland. JJS's intention to complete the above tasks resulted in his refusal to occupy Hainan Island in the latter half of 1951 at the US' request (Lin 2016: 296–297). In May 1952, JJS again refused the US' request for the ROC to attack and control Hainan Island (Lin 2016: 298). The rationale behind JJS's refusal can be understood through the history of *Bai Tuan* (the White Group).

The White Group was a clandestine military advisory team led by a retired Japanese general, TOMITA Naosuke, whose *nom de guerre* was BAI Hongliang (hence *Bai Tuan*). In February 1950, the White Group was formed and began to help JJS train ROC more than 10,000 military offices (Lin 2016: 321). The group also developed military plans and strategies for JJS, the most important one being *Guang Jihua* (the Plan of Glory), which was formulated in May 1953 (Lin 2016: 322).

JJS was very satisfied with the Plan of Glory, a military strategy he considered feasible. According to Hsiao-ting Lin (2016: 322), JJS noted that the plan was the same as what he had originally conceived. In the plan, the White Group concluded that the ROC's military forces needed another five years to complete their preparation for war against the PLA. According to the plan, the Nationalist armed forces should launch an amphibious assault to occupy the Pearl River Delta as the first step to recover the Chinese mainland. Six months after the occupation of Guangdong Province, the ROC's military forces could then proceed to take back areas south of the Yangzi River. The plan recommended that JJS prepare 1,650 aeroplanes, 298,200 tonnes of naval vessels and an army consisting of 52 infantry and 8 armoured divisions. Ultimately, the White Group believed that JJS and his armed forces would be ready to launch their counterattack in Spring 1958 (Lin 2016: 322).

Nevertheless, the Nationalists under JJS's rule did not actually cease their military operations against the Chinese Communists during the time when JJS deliberately adopted a defensive military strategy to prepare for the future offence. He issued his Surprise Attack Plan and Limited Counterattack Operation Plan (Yang 2003a: 55). He also cooperated with Western Enterprises Incorporated, which was clandestinely established and sustained by the US Central Intelligence Agency (CIA) in March 1951 (Lin 2016: 300). They initiated dozens of minor, guerrilla amphibious raids into the coastal areas of Zhejiang and Fujian Provinces between mid-1951 and mid-1953 (Lin 2016: 302).

Notably, the failure of the Dongshan Island raid in July 1953 marked the end of JJS's mainland coastal raiding operations. The heavy losses inflicted on the ROC's force (including more than 3,300 casualties, two tanks destroyed, two aircraft lost and three landing ships sunk; Lin 2016: 316) made it evident that the PLA had adequately fortified its coastal defence and naval power. Similar operations would not yield any meaningful results but would only damage the morale of the Nationalist armed forces. Furthermore, after the conclusion of the Korean War, the US strove to limit the ROC's military posture to defensive purposes only (Lin 2016).

In the above passages, I have discussed the military aspect of JJS's grand strategy before signing the Mutual Defense Treaty between the United States of America and the Republic of China (also known as the Sino-American Mutual Defense Treaty, SAMDT) in March 1955. Below, I briefly discuss the function of the ROCN in JJS's grand strategy.

Amphibious attacks were the main military strategy during the early 1950s. The most critical mission assigned to the ROCN at that time was to safely transport ROCA soldiers to the coastal areas and assist them in successfully landing at the designated places. Moreover, to transport army soldiers for amphibious landings, the ROCN had to secure the command of the Taiwan Strait.

Since the establishment of the PRC on 1 October 1949, it had made great efforts to fortify its naval power. The PLAN concentrated on building fleets consisting of a naval air force, fast boats and submarines equipped with missiles and torpedoes during the era of Mao Zedong's rule (1949–1976). To successfully transport soldiers to the coastal areas of the Chinese mainland and control the waters of the Taiwan Strait, the ROCN had to be strengthened.

The ROCN was assisted by the Navy Advisory Division (NAD) of the US Military Assistance Advisory Group (MAAG) to develop its capacity. The US MAAG arrived in Taipei on 1 May 1951. Following the advice of the NAD of the MAAG, the ROCN strove to serve the functions listed below:

> 1 to defend its own major naval bases and patrol the littoral waters along the southeast coast of mainland China; 2 to destroy all Communist forces preparing to attack; 3 to annihilate any Communist sea-borne invasion

> force; 4 to harass any Communist landing force ashore, and land and support its own counter-landing forces defending Taiwan, the Penghu islands, and forward bases.
>
> (Yang 2003b: 81)

Together with launching amphibious landing assaults, these naval functions meant that it was necessary for the ROC's military and naval forces to dominate the Taiwan Strait and secure the sailing routes of this marine area. During this period, the ROCN still enjoyed the Mahanian command of the sea vis-à-vis the PLAN.

The failure to assault Dongshan Island in July 1953 might not have revealed that the balance of naval power was tilting towards gradually favouring the PLAN. However, the sinking of the destroyer Taiping in November 1954 and the severe damage to the destroyer Taiho in January 1955 signalled that the ROC had effectively lost its command of the sea, at least in the coastal marine area of Zhejiang Province (Chi 2013: 90–91).

The destroyer Taiping was sunk by the PLAN's torpedo boat at night. This meant that the PLAN had acquired the capacity to shoot torpedoes in the dark and that it was no longer afraid of challenging the ROCN (Lin 2016: 343). The destroyer Taiho and several ROCN vessels were bombarded in Dachen Harbour in January 1955. This meant that the coastal areas of Zhejiang Province were out of bounds of the ROC Air Force (ROCAF). The PLA Air Force (PLAAF) took advantage of this geostrategic opportunity and secured the command of the air around the area outside Zhejiang Province (Chi 2013: 91).

On 18 January 1955, Mao Zedong ordered an attack on Yijiangshan Island, which is approximately nine miles north of Dachen Island (Lin 2016: 344), the base of the Nationalist guerrilla forces. Having occupied Yijiangshan Island, the PLA was poised to assault Dachen Island with the forces of its three military branches. The US, JJS and many other high-ranking military officers all understood it was necessary to evacuate Dachen Island (Lin 2016), because the ROC's military forces stationed in that area were not under adequate protection of the ROCAF and, in turn, had no guarantee of the command of the sea (Chung 2003b). Furthermore, Dachen Island did not play an important role in the defence of Taiwan. It was clear to JJS that he had to give up Dachen Island to alleviate the anxiety of the US about being dragged into war and secure the formal alliance between the ROC and the US (Lin 2016).

After the ROC's military forces evacuated Dachen Island, the ROC was still under US military assistance and made efforts to enhance its naval power. For example, US naval vessels (including 4 destroyers, 5 patrol crafts, 2 coastal minesweepers and 28 landing ships) were transferred to the ROCN. Furthermore, the US Seventh Fleet dispatched its 92nd Destroyer Squadron to the Taiwan Strait, and the ROCAF was able to gain air supremacy over the strait (Chung 2003b: 117).

Therefore, even though the US naval forces sailed the Taiwan Strait 'up to a distance of 12 nautical miles from the Chinese coast' (Chung 2003b: 117), to avoid direct military confrontation between the US Navy and the PLA, the ROCN could still confidently patrol the coastal area of Fujian Province and escort its supply convoys (Chung 2003b). Moreover, JJS also boldly increased the number of troops deployed on Kinmen and Matsu Islands from around 30,000 in 1954 to approximately 100,000 in 1956 (Lin 2016: 346). Such an increase was a part of JJS's grand strategy. JJS awaited a military conflict like the Kinmen Bombardment, which began in August 1958, to involve the US in a war directly against the Chinese Communists.

Eventually, the US did not directly participate in the conflict, and the ROC's army was strong enough to endure the bombardment. The ROCN also succeeded in resupplying Kinmen, and the ROCAF was able to retain its command of the air (Chung 2003b). Moreover, the ROCN won battles at sea and sunk many PLAN vessels (Li 2013: 105–112). Above all, according to their use and representation of the ocean, the ROC's military forces, before and during the 1958 Taiwan Strait Crisis, still constructed the ocean in the style of the original Mediterranean Sea Model.

Nevertheless, the balance of military power, especially naval capacities, between the ROC and PRC started to gradually shift from the former to the latter after the 1958 Taiwan Strait Crisis. In general, with the assistance of the US and geographical conditions, between 1959 and 1964, the ROC's military power was relatively stronger than the PRC's in the theatre of the Taiwan Strait. However, the ROCN was defeated in each of the three sea battles that took place in 1965. Moreover, following the failure of the PRC's Great Leap Forward economic policy and the resulting famine, JJS dispatched his special operation forces to infiltrate the Chinese mainland via sea lanes after 1962. But these clandestine operations ended in failure at sea (Chang 2001: 561). The ROCN was unable to accomplish its missions even after the Chinese mainland suffered from the famine. The sea battle on 6 August 1965 (the Dongshan Naval Battle) was the worst in the post-1949 history of the ROCN (Chung 2003b: 118; Xu 2002: 89–93).

With respect to naval power, after the 1958 Taiwan Strait Crisis, the PLAN embarked on a mission to convert itself from a brown-water to a green-water navy. Notably, losing three sea battles in 1965 did compel the US MAAG to help the ROCN reform and rebuild. Once again, sequestered US vessels were restored and transferred to the ROCN (Chung 2003b: 118). Nevertheless, while the fighting capacity of the PLA generally began to deteriorate during the Cultural Revolution (1966–1969) and its subsequent political factional struggles (1969–1976), JJS did not launch any military campaign to retake the Chinese mainland, mainly because the US was lukewarm about assisting any military action by the ROC against the PLA.

Another reason for the absence of ROC military actions against the PRC was the change in JJS's grand strategy after the three naval defeats in 1965. JJS accepted that the ROCN had lost its command of the sea and that it would be difficult to reclaim the Chinese mainland without the direct military involvement of the US (Xu 2002: 92–93).[2] In January 1966, JJS revised his grand strategy from an actively offensive vision into one of military defence with a diplomatic-ideological offence.

Above all, despite the Cultural Revolution's damage to the PLA's fighting capacity, the strengthening of the PLA before this fanatical political tragedy had demonstrated to JJS that it would be extremely difficult, if not impossible, for him to retake the Chinese mainland. The PLA's military enhancement forced the ROC to alter its grand strategy as well as its military forces' construction of the ocean. The many troops stationed in Kinmen and Matsu became a strategic defensive army in development. 'Naval assets, to some degree, were regarded merely, as floating supply lines' (Edmonds and Chen 2003: 95). After 1965, military battles between the ROC and PRC ceased, and there were only some minor army cannon fire exchanges in the Kinmen-Xiamen area (Chang 2001: 562).

Thus, the essence of JJS's grand strategy implicitly changed to defence. In terms of function, the ROCN was reduced to military transportation and a symbol of claims to sovereignty (for example, amphibious readiness for counterattacks against the PRC and patrolling in the South China Sea to exercise sovereignty claims over certain islands). Although controlling the Taiwan Strait was a rhetorical guideline of his military strategy, the reality was that the armed forces 'started to fortify both Taiwan and Penhu so that the ROCAF and ROCN would not have to maintain constant contact with the enemy' (Tsai 2003: 5). After 1965, the ROC's military forces gradually transformed their way of constructing the ocean from the original Mediterranean Sea Model towards that of the Indian Ocean Model. And two factors affected such a transformation: the ROC's grand strategy and the shifting balance of military (and naval) power across the Taiwan Strait.

A Brief Summary for the Final Comparison

After 1949, JJS's grand strategy, with a strong offensive orientation, consisted of four aspects whose overwhelming focus was to defeat the PLA and reclaim the Chinese mainland. Correspondingly, the military strategy of the ROC's armed forces, with the help of the White Group, aimed to achieve the grand strategy. In the early 1950s, JJS refused US' requests to retake Hainan Island, giving priority to reforming and rebuilding the military forces first.

The ROCN, then, not only played a critical role in launching amphibious assaults on the coastal areas of the Chinese mainland but also was expected to dominate the Taiwan Strait and secure the coastal marine area of Fujian Province. However, the Yijiangshan Battle was a warning to the ROC's armed forces that air supremacy was the precondition for the command of the sea. The military was quickly strengthened with US' assistance and prevailed in the 1958 Taiwan Strait Crisis. Between 1949 and 1965, the military constructed the ocean according to the original Mediterranean Sea Model.

The military balance between the ROC Armed Forces and the PLA began shifting gradually towards the latter after 1958. Many clandestine military infiltration operations conducted by the ROC's special forces and navy had failed since then, and the three nocturnal sea battles that occurred in 1965 ultimately compelled JJS to change his grand strategy from one of active offence to staunch defence. The military and naval strategies also changed as a consequence. The ROCN began to play a symbolic and secondary role in the defence of the territory that was substantially controlled by the ROC. Thus, the social construction of the ocean by the military gradually transformed into one of the Indian Ocean Model after 1965.

In the next section, I shall continue to interpret the history of the ROC's grand strategy and its armed forces' construction of the ocean between 1971 and 1991.

Grand Strategy and the Military's Construction of the Ocean between 1971 and 1991

The ROC's Grand Strategy between 1971 and 1991

With regard to the economic dimension of the ROC's grand strategy, economic development had been one of its critical pillars since it withdrew to Taiwan in 1949. Between 1971 and 1991, economic development persisted in a key role in the grand strategy. Effectively, Taiwan's economy became more anchored in the state's defensive grand strategy. To maintain the state's military power vis-à-vis the PLA, a sound economic basis was required to generate adequate revenue, as mentioned in Chapter 1. Building an industrialised capitalist economy on Taiwan also bolstered the state's legitimacy domestically. Economic prosperity, thus, assisted the defence of Taiwan.

Prior to 1971, the economy in Taiwan had already stabilised and improved, and economic policies had been implemented and adjusted. In addition to the previously mentioned economic policies that consisted of land and currency reforms, the KMT-dominated state also promoted and implemented the Reform of Foreign Exchange Trade in 1958 and the Nineteen-point Fiscal Reform Program in 1960 (Chu 2017: 275–284). These reforms successfully converted Taiwan's economy from one of import substitution into one of export orientation (Chu 2017; Wade 2004).

An export-orientation policy began to be implemented roughly in 1958, and Taiwanese exports started to grow rapidly in the 1960s. The percentage of exports contributing to the Gross Domestic Product (GDP) increased from around 12 per cent in 1962 to about 21 per cent and to approximately 48 per cent in 1972 (Chu 2017: 462). The annual growth rates of industrial production, exports and GDP in Taiwan in the 1950s were 11.9 per cent, 22.1 per cent and 8.1 per cent, respectively. These figures grew to 16.5 per cent, 26 per cent and 9.7 per cent, respectively, in the 1960s (Chu 2017: 284). The trend of rapid economic growth and state coordinated development and industrialisation continued well into the 1970s and 1980s.

The state's grand strategy began shifting to an increasingly defensive posture after 1965 in terms of military practices instead of political rhetoric. The US financial aid also ended the same year (Fell 2018: 19). In 1971, the Chinese Communists and the US reached a diplomatic rapprochement, and the UN voted for recognising the PRC as the sole legitimate representative of the whole Chinese nation on 25 October (Lai 1994: 465). On 28 February 1972, the PRC and the US signed the Shanghai Communiqué, which became a diplomatic framework for the two states to normalise and improve relations (Chung 2012: 130). Furthermore, the PLA's capacity and technology in general already surpassed that of the ROC's armed forces in 1972, as shown in the CIA's report to the US president (Lai 1994: 496).

From this time on, the geopolitical relationship between the ROC and the US was transformed from one of interdependence to one of dependence of the former on the latter (Chung 2012; Lai 1994: 455). Nevertheless, the ROC under JJS's rule continued to build an industrialised capitalist economic base in Taiwan to sustain its military fortifications and to establish legitimacy for the people in Taiwan. Its goal was to create a model province to demonstrate the superiority of the ROC's government to the people on the Chinese mainland.

JJS's son, JIANG Jingguo (also known as CHIANG Ching-kuo; hereafter, JJG) was appointed as the premier of the ROC on 1 June 1972. By November 1973, soon after the oil crisis occurred, he announced his famous Nine Major Construction Projects (later became the more widely known Ten Major Construction Projects), which he aimed to complete in five years. The purposes of these projects were to further boost the economy and to improve Taiwan's infrastructure. They included 'expanding the island's railway network, railway electrification, developing motorways and opening Taiwan's first nuclear power plant, and integrated steel plant, Taoyuan International Airport and petrochemical complexes' (Fell 2018: 19). Furthermore, to cope with the inflation caused by the energy price hike (gasoline price by 85 per cent and electricity by 80 per cent) and to maintain, and even stimulate, the economy, if possible, the ROC reduced taxes and tariffs. Thus, economic growth returned to more than 10 per cent a year in the latter half of the 1970s (Wade 2004: 97).

Led by JJG, the state in the 1980s promoted 'non-energy-intensive, non-polluting and technology-intensive activities like machine tools, semiconductors, computers, telecommunications, robotics, and biotechnology' (Wade 2004: 97–98). A series of policies developing the industrialised capitalist economy on Taiwan, as mentioned earlier in Chapter 2, were stipulated and implemented, for example, the Ten-year Plan of 1980–1989 and the Four-year Plan of 1982–1986 (Wade 2004: 98). Consequently, the average per capita income increased 'in spite of continued population growth, from US$48 in 1952 to US$4,990 in 1987' (Fell 2018: 19). The annual growth rates of the GDP of Taiwan in 1975, 1980, 1985 and 1990 were 4.9 per cent, 7.3 per cent, 4.1 per cent and 6.9 per cent, respectively, and the average GDP per capita in those years were US$978, US$2,385, US$3,292 and US$8,124, respectively (Chu 2017: 468). By the late 1980s and early 1990s, the ROC won a reputation of successful economic development and obtained legitimacy and popular support (Fell 2018: 19).

Regarding ideological aspects of the ROC's grand strategy, it functioned similarly to the economic aspects; both aimed to strengthen its internal and external defence against possible a Chinese Communist invasion of Taiwan. Although JJS passed away on 5 April 1974, the ROC still upheld the TPP. Effectively, creating a prosperous, industrialised capitalist economy on Taiwan achieved one of the main goals of the TPP. Another goal of the TPP that the ROC's grand strategy sought to achieve was the liberalisation of society and the democratisation of the state.[3] JJG expected that establishing a liberal constitutional democracy in Taiwan would generate at least two positive effects.

For the first positive effect, the liberalisation of Taiwan society became JJG's attempt to ensure the survival of the ROC, to compete against the PRC and, ultimately, to reunify Taiwan with the Chinese mainland under the guidance of the TPP. Although the ROC military posture had been defensive since 1965, the state under JJS and JJG had not publicly surrendered the goals of removing the Chinese Communists and restoring the ROC regime on the Chinese mainland. JJG made this stance clear in a conference with his high-ranking military officers on 18 November 1986. At that conference, he also pointed out the other two state objectives: state security and societal stability (Hau, 1995: 319).

The legitimacy of the state and the TIM were two threats to state security and societal stability that JJG had to handle. The challenge of sustaining state legitimacy under the KMT's authoritarian regime was alleviated by successful economic development. JJG's approach to the challenge of the TIM was to absorb more Taiwanese societal elites into the ROC regime and to incrementally liberalise society and democratise the state (Fell 2018; Hau 1995). JJG died, however, six months after he lifted martial law on 15 July 1987 (Hau 1995: 370). The then Chief of the General Staff, a close military lieutenant of JJG, HAO Bocun (also known as HAU Pei-tsun; Hau 1995: 1), points out in his memoir that JJG had long been aware of the need to establish a liberal constitutional democracy on Taiwan, and considered that JJG himself was the only one able to liberalise Taiwan society and absorb Taiwanese elites into higher positions of the state (Chung 2012: 180; Fell 2018: 27).

JJG began the liberalisation process and prepared to initiate democratisation to demonstrate to the CCP and the people on the Chinese mainland that Taiwan under the ROC's governance had put the TPP into practice. The process also demonstrated to the local Taiwanese people that it was possible to hold political power within the state and that it was also the responsibility of the Taiwanese elites to realise the TPP in the Chinese mainland and support the ROC. JJG placed his faith in the Taiwanese people (Chung 2012; Hau 1995: 317). Liberalisation and democratisation were the approaches of JJG's grand strategy to handle the threat posed by the TIM to the existence of the ROC on Taiwan.

The second positive effect of the liberalisation of Taiwan was related to the geopolitical aspect of the ROC's grand strategy during this period. The ROC needed to cope with the PRC's reunification proposal contained in *Gao Taiwan Tongbao Shu* (A Letter to the Compatriots on Taiwan), formally addressed by the Standing Committee of the National People's Congress of the PRC on 1 January 1979. The letter called for a political dialogue across the Taiwan Strait and the establishment of *San Tong* (the Three Links): *Tong You, Tong Shang, Tong Hang* (Establishing Direct Postal, Trade and Transportation Links) between Taiwan Island and the Chinese mainland. DENG Xiaoping, the then prime leader of the PRC, soon afterwards also 'made clear that the concept of "one country, two systems" adopted for Hong Kong and Macau was also designed for Taiwan' (Fell 2018: 176).

In response, JJG issued his *San Bu Zhengce* (Three Nos Policy; No Contact, No Negotiation and No Compromise) in April 1979. The Three Nos Policy was the official stance of the state. However, in October 1987, JJG adopted a new, flexible policy allowing people on Taiwan to travel back to the Chinese mainland for family visits (Chung 2012: 188; Fell 2018: 177). In terms of grand strategy, JJG's intentions were twofold.

First, he expected that the connections and transactions of civilians across the Taiwan Strait would prevent the dilution of Chinese culture and would strengthen Chinese national identity in Taiwan. Furthermore, JJG believed that it was his duty to take care of the veterans who had followed his father in the state's withdrawal to Taiwan and that by allowing them to return to the Chinese mainland to visit their families, he not only fulfilled his duty but also enhanced the influence of the KMT version of Chinese nationalism and restricted the effect of the TIM (Chung 2012: 188; Hau 1995).

Second, by the mid-1980s, JJG was confident that the gradual political liberalisation and successful economic development of Taiwan could be effective instruments to initiate social change on the Chinese mainland. Moreover, he strongly believed that the Chinese Communist regime had to transform itself and its governance to catch up with the quality of the ROC's regime and Taiwan's economic achievements (Taylor 2000).

Another major political issue that the state's grand strategy had to deal with was the question of arms sales by the US, JJG had countered the US demands for Taiwan's democratisation through by liberalising society since the mid-1970s (Fell 2018: 37). In terms of diplomatic image, liberalisation increased support for the ROC by the US Congress and other democratic states. More importantly, early in 1976, JJG had begun lobbying the US Congress to prepare for the worst scenario of the ROC-US relationship: the diplomatic derecognition of the ROC and termination of the SAMDT, which eventually happened on 1 January 1979 (Fell 2018: 176).

Soon after the US formally recognised the PRC as the sovereign representative of the entire Chinese nation (including the population on Taiwan), the

US Congress passed the Taiwan Relations Act (TRA) by the end of March 1979 (Chung 2012: 164). Although 17 August Communiqué between the PRC and the US restrained both the quantity and quality of US arms sales to Taiwan, the ROC could still acquire US defensive weapons to meet its security needs according to the TRA (Hau 1995: 68–74). Additionally, because it became more challenging to obtain US weapons, the ROC under JJG's rule began to purchase arms from European countries and to produce needed weapons domestically (Hau, 2000). The arms trade and domestic arms production became the key to the modernisation of the ROC's military forces, which, in turn, was a core aspect of the military strategy during this period.

Above all, the ROC under the administration of JJG continued to improve and rapidly developed Taiwan's economy. The economic dimension of the state's grand strategy was to develop an industrialised capitalist economy. The liberalisation of the society on Taiwan and the democratisation of the ROC, as well as economic prosperity, were the main pillars of the TPP. Following the prescription of the TPP, JJG planned to promote liberal constitutional democracy in Taiwan; this plan constituted the ideological core of the state's grand strategy, even though liberalisation was incremental, and democratisation was embryonic prior to JJG's death.

The approach to incremental liberalisation and embryonic democratisation was expected to assimilate more Taiwanese to support the ROC and its mission of restoring sovereignty over the Chinese mainland. Geopolitically, this approach alleviated the US pressure for total liberalisation and full-fledged democratisation. Moreover, together with Taiwan's economic performance, partial liberalisation and democratisation became the best political weapons against the Chinese Communists. Furthermore, JJG also secured the stipulation of the TRA by the US Congress. JJG's diplomacy enabled the ROC to continue acquiring US weapons for defensive purposes via the legal channel of the TRA.

By now, it had become clear that the ROC under JJG's rule had adopted a staunchly defensive grand strategy. Its offensive element was limited to the diplomatic and ideological arenas. Against this background, I will discuss how the ROC's military forces constructed the ocean between 1971 and 1991.

The Construction of the Ocean by the ROC's Military Forces between 1971 and 1991

Nuclear weapons can easily overwhelm and nullify defence with conventional armed forces. The PRC gained nuclear weapons and related delivery capability in the second half of the 1960s. Although the PRC has maintained a no-first-use policy for its nuclear weapons since the 1960s (You 1999: 86), JJG and the ROC Armed Forces still had to deal with the PRC's nuclear capacity. Soon after the PRC's successful nuclear test, JJS responded by establishing the National Chung-Shan Institute of Science and Technology in 1965 for secretive nuclear weapon development. However, in 1975, JJG publicly acknowledged the ROC's efforts to produce nuclear weapons and renounced any plan to use or further develop the nuclear capability (Chung 2012: 200).

Strategically speaking, JJG's renunciation made any possible use of nuclear weapons by the PLA against the ROC and Taiwan society illegitimate and unjustifiable. The discourse that JJG utilised to explicate the reason for abandoning nuclear aspirations was nationalistically sentimental, exemplified by JJG's speech to the Legislative Yuan (the law-making branch of the central government of the ROC):

> After 17 years of effort ... we now have both the facilities and the capability to make nuclear weapons and actually considered building up a nuclear arsenal last year [1974]; but I broached the idea to late president [JJS], he rejected it flatly on the political ground that we cannot make nuclear weapons to kill our countrymen.
>
> (Chung 2012: 200–201)

Deploying a public discourse that renounced the use of nuclear weapons against the PRC on the grounds that people on both sides of the Taiwan Strait belonged to the same nation and, thus, weapons of mass destruction should not be wielded against each other, allowed the ROC to occupy the moral high ground and forestalled any justification by the PRC for using nuclear weapons against Taiwan.

In addition to his strategy for responding to the PLA's nuclear weapons, JJG also had to deal with the PLA's conventional forces. A starting point for the discussion is a comparison of conventional weapon capacity. In 1984, the PLAN owned 103 submarines, 11 destroyers, 34 corvettes, 904 gunboats, 23 minesweepers, 61 large amphibious landing ships (plus 450 small landing boats) and 44 other auxiliary naval vessels. The ROCN, for its part, possessed 2 submarines, 20 destroyers, 11 corvettes, 44 gunboats, 22 minesweepers, 33 sizeable amphibious landing ships (plus 270 small landing boats) and 28 other auxiliary naval vessels. The ROCN used *Hsiung Feng* (Strong Wind), Gabriel, Sidewinder and Sea Chaparral missiles, which were qualitatively superior to the SSN2 used by the PLAN at the time. The PLAAF kept around 4,000 fighter aircraft in the same year, whereas the ROCAF maintained approximately 500. Regarding air-to-air missiles, the PLAAF utilised missiles of the PL series. The ROCAF employed Sidewinders with better quality (Chang 2001: 570–571).

With regard to the size of the overall military forces of the ROC and that of the PRC, the former had maintained a constant size of around 480,000 between the mid-1970s and 1991 (Chang 2001: 566). The size of the PRC forces began to decline from about 6.1 million in 1975 to roughly 4.5 million in 1978, and by the mid-1980s, it had decreased to 3.5 million. From the mid-1970s to 1989, the PLA underwent a process of reorganisation and reduction of scale (You 1999: 33), although its size was still much larger than that of the ROC.

Although the PLA retained its traditional military strategic 'guidelines: waging campaigns through the manoeuvre of large masses of soldiers in offence and deploying fixed positional warfare in defence' (You 1999: 2), due to its low technological level in the 1970s and 1980s, its sheer size still posed an immense threat to the ROC and made it impossible for the ROC to consider re-establishing itself on the Chinese mainland without direct US military involvement. Thus, it is understandable why JJG, as the vice-premier of the time, instructed the chief of the general staff LAI Mingtang on 27 October 1971 (the second day after the ROC forfeited its seat in the UN) that 'from now on, all our military preparations should be based on a defensive (*fang yu*) posture. There is no need to make much preparation for reunification from now on' (Lai 1994: 470, cited in Chung 2012: 194).

On 16 March 1982, then Legislator HUANG Huang-hsiung in the interpellation of the entire body of the Legislative Yuan questioned the direction of the military strategy. The then minister of national defence, SONG Changzhi confirmed that the contemporary military strategy was one of strategic defence (Huang 2017: 43). Later, the then premier, SUN Yuanxuan supplemented the explanation by claiming that the ROC adopted a grand strategy that had been militarily defensive and politically offensive (Huang 2017: 44).

That occasion was the first time the ROC had publicly acknowledged that the grand strategy was a defensive one. Moreover, JJG also openly acknowledged the strategies on 16 May 1983 (Huang 2017: 45). Facing the Chinese Communist threat with no military advantage, JJG had to increase defence spending and promote military modernisation is an effort to raise the quality of the ROC Armed Forces and to balance the military power with that of the PRC. Military modernisation was the critical component in the ROC's grand strategy. Therefore, informing the national legislature of the grand and military strategies was a friendly gesture to secure an adequate defence budget while minimising resistance. JJG strove to modernise the armed forces through two approaches: foreign arms procurement and domestic weapon development (Hau 2000).

In the 1970s and 1980s, the PLAN could be viewed as a green-water navy with the ability to threaten the sea lanes of communication (SLOCs) along the coast of the Chinese mainland and the South China Sea (Chung 2003b: 118). The ROC's military during this period determined that it would be unwise to dispatch naval vessels to areas where air supremacy was not secured. This idea was derived from the ROCN's own experience in the mid-1960s and during the 1982 Falklands War (Hau 2000).

Additionally, even though the ROC Armed Forces still maintained higher quality weapons than those of the PLA, the ROCN and ROCAF remained vulnerable to the PLA's saturation attack from the coast of the Chinese mainland and from the air. Therefore, the ROCN 'quietly modified its mission and placed the protection of its SLOC and waters around the main island of Taiwan as its top priority instead of fighting a war around the offshore islands' (Chung 2003b: 120).

When the defence focused on the main island of Taiwan, the ROC's forces had to deal with a possible PLA naval blockade, which could be conducted with naval mines, submarines, or even naval surface vessels (Hau 2000: 87). Thus, the ROC's military forces attempted to strengthen their capabilities through anti-submarine aircraft and naval minesweepers. Drawing lessons from the sinking of the British *HMS Sheffield* by a French-built Exocet anti-ship missile launched from an Argentinian fighter aircraft, HAO Bocun asserted that large-scale ships equipped with ship-to-ship missiles were not suited for the ROC's military forces.

HAO Bocun suggested that 'when air supremacy is secured to allow the use of air-to-ship missiles, ship-to-ship missiles are not needed, and large naval battleships equipping missiles are outdated. The command of the sea can be achieved with small-scale naval vessels and submarines' (Hau 2000: 100). Instead of relying on naval warships for anti-blockade missions, he built the ROCN as a green-water navy supported by land-based ROCAF. In the case of Taiwan, he believed that the command of the sea had to be secured from the air and from the land (Hau 2000).

In response to the danger of a PLA saturation attack against the ROCN and ROCAF, JJG and HAO Bocun utilised the high mountainous terrain of Taiwan and fortified the underground Jiashan Air Force Base underneath the eastern side of the Central Mountain Range. Moreover, at the eastern coast of Taiwan, Suao Port was expanded and upgraded during JJG's Ten Major Construction Projects. The Suao Port was used as the SLOC terminal with the US when the ports along the western coast of Taiwan were blockaded by the PLA. The ROC's military forces were assigned the joint responsibility of guaranteeing the security of the eastern SLOC of Taiwan.

Amongst the three branches of the armed forces, the ROCAF received the most resources and was ordered to provide vital air support during the final decisive battles on the shores of Taiwan, while the ROCA was deployed to offer land-based missile defence to protect the ROCAF and ROCN (Edmonds and Chen 2003: 98; Hau 2000). The ROCA also had to repel the PLA's landing troops. The ROCN fleets 'should, therefore, evacuate to the Pacific Ocean, either to secure Taiwan's eastward "safety route" (the possible sea route for foreign aid), or to wait in some safety before intercepting the PLA's amphibious forces and follow-on logistic support convoys that approached Taiwan's western shores' (Edmonds and Chen 2003: 98).

JJG and his trusted military lieutenant HAO Bocun understood that the final stage of the PLA's invasion of Taiwan would be an amphibious landing. Because they believed that the decisive battles would be on the shores of Taiwan, the army was still considered the key to the final survival. The ROCA was designed to be a fast-striking force capable of defeating PLA units within a 20-kilometre band between the shore and the interior and of swiftly regrouping during the interval of the different waves of amphibious landing attempts (Hau 2000: 126–127, 153).

Furthermore, the ROC's military strategy emphasised the principle *Zhanlüe Chijiu, Zhanshu Sujue* (Strategic Endurance, Tactical Swiftness). 'Strategic Endurance' meant that the ROC forces had to be able to bear the PLA's first strike and maintain the capacity to counterattack in the decisive battles. 'Tactical Swiftness' referred to rapid deployments and redeployments, small and cautious skirmishes to paralyse or disturb the PLA's fighting pact and capacity and the ability to attack the weakness and critical parts of the PLA to minimise losses and maximise gains for the definitive triumph.

The above discussion about the ROC's grand and military strategies provides insight into the military's construction of the ocean between 1971 and 1991 through the Steinbergist Classification. After 1971, the ROC's grand strategy became defensive. The state governed by JJG also promoted the gradual liberalisation of Taiwan's society and prepared for establishing a constitutional democracy, even though the most effective democratisation of the state did not commence before 1996.

The offensive components of the ROC's grand strategy between 1971 and 1991 were limited to the ideological and political spheres. The military posture was transformed into a purely defensive one, concentrating the armed forces on protecting the main island of Taiwan. The nonmilitary aspects of the grand strategy constructed the ocean in a rationale similar to the overall construction of the ocean in the era of industrial capitalism in the west, as depicted by Steinberg (2001). The ocean in the industrial capitalist construction was a void across which capital flows, production and reproduction could take place. The construction of the ocean followed the style of the Indian Ocean Model, as Steinberg (2001) indicates.

The ROC's military strategy during this period, however, did not align with Steinberg's (2001) argument about the construction of the ocean by the Western navies in the era of industrial capitalism. The ROC's military forces still constructed the ocean in the manner of the Indian Ocean Model. As mentioned earlier, Steinberg (2001) believed that the Western navies constructed the ocean in the fashion of the original Mediterranean Sea Model because of their need to justify their relevance and to secure an adequate defence budget. The ROC's military between 1971 and 1991 viewed the ocean as a source of danger; it was an access route through which the PLA could invade Taiwan. Nevertheless, to avoid being annihilated on the sea by a PLA saturation attack, it was imperative that the ocean not be dominated by the ROCN and ROCAF. The enemy had to be defeated on the shores of Taiwan. The Pacific Ocean east of Taiwan became a space for the ROC and its military forces to seek US aid.

A Brief Summary for the Final Comparison

The ROC's grand strategy and the changing balance of military power between the ROC and the PRC are summarised in this section. The ROC's grand strategy between 1971 and 1991 relied heavily on Taiwan's economic performance. The ROC strove to coordinate and facilitate the development and industrialisation of Taiwan's economy. Wade's (2004) notion of governing the market summarises the economic aspect of the ROC's grand strategy. The famous Ten Major Construction Projects were completed. A series of industrial policies were implemented.

The incremental liberalisation of the society on Taiwan and the incipient democratisation of the ROC were the ideological dimensions of the state's grand strategy. Liberalisation was expected to ensure the survival of the ROC and the ultimate reunification of Taiwan and the Chinese mainland. With the plan to democratise the state and the appointment of more Taiwanese elites to important positions of the regime, JJG hoped to weaken the TIM and that Taiwanese elites would follow the TPP and support the reunification of Taiwan with the Chinese mainland under the banner of the ROC.

Liberalisation and democratisation also played an important geopolitical role in the ROC's grand strategy. JJG was confident that the liberal democracy of the ROC would force the democratisation of the PRC regime. This was one of JJG's methods to counter Deng Xiaoping's 'one country, two systems' proposal for China's reunification. JJG aimed to ultimately establish a unified China under a liberal, democratised ROC. Additionally, a liberalised society with a gradual pace of reform and a democratised ROC in its early stage could help secure the sale of crucial US weapons to Taiwan.

The military balance shifted towards the PRC after 1958, as mentioned earlier. Sea battles in 1965 clearly demonstrated PRC military superiority, and the ROC changed its grand strategy from one of active offence to defence. Although cross-strait confrontations became mainly diplomatic and ideological, the ROC still had to defend itself militarily. The ROC aimed to forestall the PLA's legitimate use of nuclear weapons against Taiwan by renouncing the development and employment of nuclear weapons against the Chinese mainland. The ROC's military forces were then dovetailed to cope with the PLA's strategy of saturation attacks by conventional forces. The modernisation of the ROC's armed forces was a critical approach. Another approach was to avoid direct military clashes in the Taiwan Strait. The site of the final military showdown was expected to be on the shores of Taiwan. Much emphasis was placed on the ROCAF to secure air supremacy; and the ROCN was ordered to intercept the PLA's amphibious forces and its logistic supply convoys. Thus, for the ROC's military forces, the ocean was viewed as a source of military threat and potential invasion. The military function provided by the ROCN, according to HAO Bocun's design, was secondary, supportive and indirect. Different from Steinberg's (2001) analysis of the construction of the ocean by modern Western navies, the ROC's military forces (including the ROCN) between 1971 and 1991 constructed the ocean in the style of the Indian Ocean Model.

Grand Strategy and the Military's Construction of the Ocean between 1991 and 2016

The ROC's Grand Strategy between 1991 and 2016

If a state's security objectives that are expected to be achieved through military, economic, ideological and geopolitical approaches can be considered a grand strategy, as defined at the beginning of this chapter, then the geopolitical aspect of the ROC's grand strategy consisted of the democratisation and democratic consolidation and the ideological aspect involved the state's promotion of the TIM between 1991 and 2016. If the plan to kick-start democratisation was part of the ROC's grand strategy under JJG's rule, during the years between 1991 and 1996, LI Denghui (JJG's successor, also known as LEE Teng-hui; hereafter, LDH) actually took actions to cultivate the seeds of democracy, making democratisation of the state possible (Jacobs 2016). Democratisation developed steadily towards democratic consolidation, whereas the promotion of the TIM was discontinued by the KMT-nominated president, MA Yingjiu (also known as MA Ying-jeou; hereafter, MYJ), between 2008 and 2016, even though the TIM was once again encouraged by the state after the DPP-nominated CAI Yingwen (also known as TSAI Ing-wen) inaugurated her presidency in May 2016. This matter is beyond the scope of this monograph.

In June 1986, the DPP was allowed to be established. In July 1987, martial law was lifted 38 years after it was imposed, and JJG sowed the seeds of democracy in Taiwan. However, JJG's sudden death in January 1988 rendered LDH a weak acting president and party chairman of the KMT, as LDH himself admits (Lee 2015). JJG did not indicate a clear path forward for the ROC's future grand strategy or for a successor of the ultimate political power. Consequently, LDH did not have any powerful allies and lieutenants to support his administration, at least in his own perception (Lee 2015: 89).

JJG's unexpected demise contributed to the fractionalisation and the eventual fragmentation of the KMT regime. The KMT had never really eliminated factional politics since the founding father of the KMT and the ROC, SUN Yixian (also known as SUN Yat-sen), suddenly lost his life to cancer. It was firmly dominated by JJS and JJG only after 1949. LDH became the acting president because he was the vice president when JJG passed away. LDH's situation and personality were not like YAN Jiagan's. Yan was the vice president when JJS died. Yan's acting presidency was expected to facilitate JJG's ascendency to power. However, precisely because JJG did not choose a successor, his unexpected departure and LDH's acting presidency provided LDH the necessary power and a structural space for political manipulations in the factionalised KMT.

Some of JJG's important lieutenants competed against each other after his death, and LDH 'proved adept at manoeuvring' (Jacobs 2016: 52). He first cooperated with YU Guohua, the then premier, 'when Li became president.

Yu had kinship links to the Chiang family and a close relationship with Mme Chiang Kai-shek' (Jacobs 2016: 52). This move helped LDH release HAO Bocun from the position of chief of the general staff and appoint Hao as the minister of national defence. LDH argues that Hao's long tenure as the chief of the general staff (from 1981 to 1989) conferred on him tremendous power over the military forces (Lee 2015: 91). LDH expected that the elimination of Hao's direct command of the ROC's military would gradually place the armed forces under the control of a civilian president rather than military personnel with KMT membership.

LDH also collaborated with LI Huan, the then Secretary-General of the KMT, to secure his KMT chairmanship. When the time was ripe, YU Guohua was replaced with LI Huan. LI Huan's control over the KMT was then transferred to LDH supporter SONG Chuyu in June 1989. When LI Huan openly vied for the vice-presidency against LDH's wish in February 1990, LDH formed a political alliance with HAO Bocun and appointed him as the new premier. Hao's formal power over the ROC Armed Forces was further curtailed when he was promoted to the premiership in June 1990 (Lee 2015: 98).

It is still uncertain whether or not LDH deliberately intended to thwart the societal momentum towards democratisation to defeat his political enemies. It is, however, undeniable that such a political scenario did unfold under LDH's rule. On 16 March 1990, when LI Huan challenged LDH's authority, 6,000 students launched what they called the 'Wild Lily Student Movement'. They 'criticised the National Assembly as well as the Temporary Provisions to the Constitutions implemented under Chiang Kai-shek' (Jacobs 2016: 95). On 20 March, LDH accepted the students' demand for the convention of a National Affairs Conference in June 1990 to accelerate the preparation for the actual transition to democratisation (Lee 2015: 96, 99).

LDH then persuaded the original central parliamentarians, who had been elected on the Chinese mainland before 1949, to cooperate and retire. Following the decisions of the National Affairs Conference, the Temporary Provisions was abolished in May 1991, and the entire members of the National Assembly were re-elected by the end of the same year (Lee 2015: 100). Furthermore, on 19 December 1992, 'Taiwan's voters went to the polls to choose the first legislature completely elected by Taiwan's voters' (Jacobs 2016: 55).

DENG Xiaoping's *Muo zhe Shitou Guo He* (Crossing the River by Touching the Stones) appropriately describes the approach of the CCP's economic reforms after 1979. It is also well-suited to describe LDH's preparation for the democratisation of the ROC in the early 1990s, especially when it was interlinked with his political struggles within the KMT regime. Not only did LDH manipulate his political enemies (and allies) with intrigues and divert hostile societal energy for his own use (for both the consolidation of his presidential power and, according to his own testament, the democratisation and indigenisation of the ROC; Lee 2015), but he also knew where and when he should take one step back and where and when to compromise and move two steps forward.

For example, as LDH proclaims (Lee 2015: 101), he established the National Unification Council in October 1990 and stipulated the Guidelines for National Unification in 1991 to alleviate the opposition by 'the KMT old guard' (Wu 2016: 395). He claimed that the guidelines were, effectively, used by him to curb and contain any intention and momentum towards reunification (Lee 2015: 102). Additionally, although there is no concrete evidence to prove that LDH collaborated with the opposition that was demanding democracy and embracing Taiwanese nationalism to oust then premier, HAO Bocun, it was clear that, as KMT chairman and comrade, LDH did not muster his political strength to defend the premier who he had appointed. LDH, in effect, appointed LIU Heqian, a retired admiral and life-long political rival of Hao, as the chief of the general staff in November 1991, to eradicate Hao's influence over the ROC's military.

LDH was skilful at connecting his efforts to consolidate presidential power with societal demands for democratisation and the actual preparation for the transition to democracy and tactful at incorporating and cultivating the drive of the TIM. When LDH perceived that he was powerless when he first became acting president in 1988, he did not adopt a heavy-handed approach to suppress the DPP and the TIM to demonstrate his determination to protect the ROC and worthiness to lead the KMT before the powerful senior party cadres. Rather, when he convened the National Affairs Conference, many people supporting the TIM, for example, HUANG Xinjie and ZHANG Junhong, were invited.

Furthermore, if he is forthright in his autobiography (Lee 2015: 104), LDH's decision to establish the institution of a directly elected presidency by popular vote not only increased his own presidential power but also successfully communicated the message that people on Taiwan could elect their own presidents, instead of letting the KMT-controlled National Assembly select who would lead the country. The direct popular election of the presidency became a symbol, and critical element, of the discourse of Taiwanese nationalism.

LDH consolidated his political power after the second constitutional amendment that established the direct popular election of presidents passed on 27 May 1992 (Jacobs 2016: 54). Since 1993, LDH had intended to further enhance his power by connecting the state with the TIM. The first Taiwanese Secretary-General, XU Shuide, was appointed by LDH in March 1993, and XU inducted many Taiwanese into crucial positions of the KMT in an attempt to garner more support from Taiwanese society and marginalise the so-called conservatives and old guard of the KMT (Jacobs 2016: 55). In 1994, he openly described the KMT as a foreign regime in an interview with Japanese writer SHIBA Ryotaro. Furthermore, LDH's ostentatious support for the Guidelines of National Unification was deemed deceitful by the Chinese Communist leaders, given that he also undertook his *Wushi Waijiao*

(Pragmatic Diplomacy), which aimed to 'establish diplomatic ties with some African, Central American and Caribbean countries' (Huang 2016: 465). LDH also called for the ROC's return to the UN (Huang 2016: 466). The Pragmatic Diplomacy platform culminated in his visit to Cornell University in June 1995 (Hughes 2016: 158–159).

At the same time, LDH also launched his de-Sinicization campaign in school education. The importance and proportions of the education materials about China were downgraded and reduced in the new curriculum. His campaign of de-Sinicization was viewed by many senior KMT members as a type of propaganda trying to justify his 'policy to privilege the position of a new Taiwanese political elite and legitimize the concentration of power in the hands of Lee Teng-hui' (Hughes 2016: 159). As the first Taiwanese president, LDH projected an image of promoting democratisation for the people in Taiwan, especially before he won the first direct popular election of the presidency with 54 per cent of the vote in March 1996 (Jacobs 2016: 57).

In 1997, he mustered sufficient political strength to suspend the functions of the Taiwan Provincial Government, which ended 'the terms of the Taiwan Provincial Governor and the members of the Taiwan Provincial Assembly on 20 December 1998' (Jacobs 2016: 58). SONG Chuyu, a senior mainlander KMT member and the then popularly elected provincial governor who had cooperated with LDH, was uprooted from his power base. LDH once again severely damaged the power of his political opponents. Moreover, this move served an important ideological function, representing the termination of the ROC's nominal governance of the other provinces on the Chinese mainland, and became a discourse supporting the TIM (Jacobs 2016: 160).

LDH's promotion of the TIM reached its zenith in July 1999, when he claimed that the 1991 constitutional amendments had transferred the cross-Strait situation into a special state-to-state relationship, instead of a domestic relationship between a legitimate government and a rebellious group, or between a central government and a local government (Su 2009: 52). This claim has been labelled the Two-States Theory, announced because 'Lee wanted to use his power to change the status quo. Furthermore, he hoped to use the new definitions of Taiwan's status and of cross-strait relations to establish "Lee Teng-hui's path" for his successors' (Su 2009: 56).

Therefore, between 1991 and 2000 (or even earlier, from 1988), the consolidation of the power of the presidency became a part of the state's grand strategy and was intertwined with the ideological aspect of the strategy. In the beginning, political alignments and positional reshuffling were employed by LDH. At the same time, the societal forces demanding the democratisation of the state were co-opted by LDH's regime. This constituted the geopolitical aspect of the ROC's grand strategy. Afterwards, the state became connected to the TIM, and LDH made the promotion of the TIM the primary focus of the ideological dimension of the grand strategy.

The democratisation of the ROC certainly went beyond being a strategic element of the state. Although democratisation cannot be viewed as a mere product of the state, the state promoted and sustained the impetus of democratisation when the incumbent followed the constitutional requirements to transfer the controls of state power to the opposition. In 2000, the KMT lost its control of the ROC for the first time since the end of 1928 to the DPP. CHEN Shuibian (also known as CHEN Shui-bian; hereafter, CSB) became the new president in May of that year. The transfer was both peaceful and constitutional. The democratisation of the ROC gained its own momentum and, from then on, moved steadily towards democratic consolidation, despite the domestic and international questions about the quality of liberal democracy on Taiwan after 2000 (Fell 2018: 44).

While democracy had been established and was gradually moving towards consolidation, the ideological aspect of the state's grand strategy became radicalised after 2000. In the first year of his first term, CSB appeared to show his intention to reduce the tension between the two sides across the Taiwan Strait (Su 2009). He even mentioned that he would follow the *Si Bu Yi Meiyou* (Four Nos and One Have Not, also known as the Five Nos) principle, which means 'no change of the national name, no change of the ROC constitution that would include Lee Teng-hui's "state-to-state theory", no referendum on independence, no declaration of [Taiwan Independence,] and no intention to abolish the Guidelines for National Unification and the National Unification Council' (Hughes 2016: 160). However, in May 2001, the cross-Strait relationship turned sour and began to deteriorate again (Su 2009: 139).

SU Chi (2009), former Secretary-General of Taiwan's National Security Council, legislator in the Legislative Yuan and cochair of the National Defense Committee of the Legislative Yuan, provided a first-hand observation and explanation for CSB's failure to improve the cross-Strait relationship in the first year of his presidency. He points out that the window of opportunity to improve the relationship 'was lost because other than the publicly known "five Nos", announced in the inauguration speech there was a hidden and more influential set of "five Nos"' (Su 2009: 129).

The first hidden 'No' was no One-China consensus. CSB's vice-chairman of the Mainland Affair Council, CHEN Mingtong and the general secretary of the Strait Exchange Foundation, XU Huiyou, publicly denied that the ROC during LDH's presidency had reached a consensus with the CCP about the One-China issue. CSB's vice-president, LÜ Xiulian, even proclaimed that Taiwan's sovereignty was undetermined according to international law. Lü claimed that Taiwan had not been legally occupied by the ROC after Japan lost the Second World War and surrendered. CSB's chairwoman of the Mainland Affairs Council, CAI Yingwen, also denied the One-China consensus immediately after CSB publicly mentioned that he might accept the One-China consensus (Su 2009: 130).

'The second hidden "No" was "no positive statements concerning any high level issue"'. Typically, his administration expressed its views only in the negative on those issues of great concern to the international community and the PRC such as 'sovereignty, DPP objectives, and principles' (Su 2009:

131–132). The third hidden 'No' was that CSB made no efforts to formulate the intra-DPP consensus on the cross-Strait relationship. The cross-Strait relationship was also a divisive issue within the DPP. Without such an intra-party consensus, it would not have been possible for CSB to adopt a coherent stance and potentially make compromises with the PRC to ameliorate the cross-Strait relationship (Su 2009).

The fourth hidden 'No' was no public contact or negotiation with the PRC. This was because CSB agreed that 'open official contact and negotiations between the two sides of the Strait were to be avoided at all costs (secret contact was another matter), lest Taiwan forgot who its enemy was' (Su 2009: 136). The last hidden 'No' was no amicable reply to the CCP's friendly statements and gestures. When the CCP leaders delivered speeches that emphasised the fact that the Chinese mainland and Taiwan are both parts of China, shifting the CCP's stance from simply claiming Taiwan as a part of China, 'which connotes a dominant/subservient relationship', CSB refused to respond positively but with 'extreme indifference' (Su 2009: 137).

Above all, the 'public set of Five Nos declared during Chen's inauguration served to bring temporary stability to the Taiwan Strait' (Su 2009: 138). However, the concealed Five Nos made it impossible for the ROC under the leadership of CSB to reach détente with the CCP. The way to increase his own political credibility by improving the cross-Strait relationship was also obstructed. When other political figures of the TIM openly challenged his public Five Nos, Chen had to react and support the TIM with even more drastic measures, resulting in the radicalisation of the ideological aspect of the ROC's grand strategy.

When an actively pro-TIM political party, the Taiwan Solidarity Union, was established in August 2001 and won 13 seats in the election of the Legislative Yuan in December, CSB had to rely on its support to stipulate laws. Moreover, when CSB became the party chairman of the DPP, pressed by the more radical forces supporting the TIM both in and out his party, he proclaimed that 'the relationship between the two sides was that of 'one country on each side of the Taiwan Strait' (Hughes 2016: 162).

CSB, following LDH's approach, connected the state to the TIM. In addition to advocating the referendum to decide if Taiwan should apply for membership to the UN, he also implemented policies to strengthen Taiwanese national identity. 'Particularly controversial for Chinese nationalists were steps such as the removal of the name "China" from buildings and institutions of national significance and the adding of "Taiwan" to official documents, including the ROC passport' (Hughes 2016: 162). Chinese nationalism that the KMT regime used to indoctrinate the military forces was removed. In CSB's second term of presidency, he continued to resort to Taiwanese nationalism to derive support from the TIM (Hughes 2016).

The US was positive regarding the democratisation of the ROC in 1996 and the peaceful transfer of power to the opposition in 2000. The Clinton administration protected Taiwan with two aircraft carriers in 1996 and made a statement that stressed the US objection to the use of military force to resolve the Taiwan question in 2000 (Su 2009: 123). The policy orientation of the Bush administration was even more advantageous to Taiwan.

However, the September 11 terrorist attack became a watershed moment in the PRC-US relationship. 'Once "anti-terrorism" provided both sides with a strategic basis, the US-China relationship immediately acquired an impetus to improve' (Su 2009: 150). To concentrate on its priorities in the Middle East, the US 'did not want Taiwan to become a "troublemaker" again, and carelessly drag the United States into a conflict with China, with the resultant destruction of East Asian peace and stability' (Su 2009: 152). The aforementioned proclamation by CSB—that is, 'one country on each side of the Taiwan Strait'—led to the deep dissatisfaction of the US with his attitude. After August 2002, the US sent direct and clear messages to Taiwan, warning against any measures leading towards the realisation of Taiwan's *de jure* independence (Su 2009: 202).

Incidentally, CSB's radicalisation of the geopolitical and ideological aspects of the ROC's grand strategy occurred when the Taiwanese economy began to plummet. The growth rate of the economy in the first quarter of 2000 (that is, the last quarter before the KMT regime transferred state power to the DPP) was 7.94 per cent, and the second and third quarters remained at 5.1 per cent and 6.73 per cent. However, the sudden cancellation by CSB of the construction of the Fourth Nuclear Power Plant, intended to secure and stabilise the electricity supply to meet the development demands for the foreseeable future, resulted in a fall in the fourth quarter growth rate to 3.82 per cent. By 2001, it had fallen to −2.18 per cent (Su 2009: 164–165).

'Private investment – the power behind economic growth – grew at double digits every year throughout the 1990s. In 2000, Taiwan maintained a rate of growth of 15.74%, but plummeted to −29.2% in 2001' (Su 2009: 165). It is also worth noting that Taiwan's trade surpluses 'with the PRC in 2000, 2001 and 2002 stood at US$19.92, US$18.16 and US$25.11 billion, respectively. … Without the mainland's market, Taiwan's foreign trade would show a negative balance' (Su 2009: 166).

LDH and CSB believed that if they wanted to sustain the connections of the ROC to the TIM, they would have to reduce Taiwan's reliance on the Chinese mainland market. Concerned that Taiwan's dependence on the mainland market would endanger Taiwan's economy and security, both LDH and CSB sought to replace this market with access to an increasingly globalised world. Neoliberal economics also pervaded Taiwan's economic policymaking after the early 1990s, with deregulation of Taiwan's economy becoming the dominant paradigm. Financial and important infrastructural services (such as oil, electricity and high-speed railway) were privatised (Wang, Lee and Chen 2017: 53, 67).

Furthermore, LDH declared his plan to transform Taiwan into the Asia-Pacific Operation Center, as mentioned previously in Chapter 2, and CSB launched his Two Trillion and Twin Stars Program.[4] Although these projects were criticised and difficult to demonstrate their success or failure (Lee and Lin 2017; Wang, Lee and Chen 2017), they represented attempts to prepare Taiwan's economy for global economic competition and to achieve 'globalization without China' (Leng 2016: 235). This is also one of the rationales behind Taiwan's endeavour to join the World Trade Organization in 2001 (Lee 1999; Lee and Chu 2016: 413).

The most prominent features of their efforts to globalise Taiwan's economy without China can be observed in LDH's *Jie Ji Yong Ren* (No Haste, Be patient) policy and CSB's *Jiji Kaifang, Youxiao Guanli* (Active Liberalisation, Effective Management). These were designed to curb Taiwan's capitalists from investing in the Chinese mainland and to diversify Taiwan's exports to the global market. However, in terms of investment, 'Taiwan's investments in China as a percentage of its total outbound investment increased from 27.7% in 1999 to 60.6% in 2007' (Lee and Chu 2016: 411). In terms of exports, in 1999, 17.2% of Taiwanese exports went to China, and by 2014, that proportion had increased to nearly 30%' (Lee and Chu 2016: 411). Above all, from Leng's (2016) perspective, the globalisation of the economy on Taiwan without China was the economic aspect of the ROC grand strategy under LDH and CSB.

The 2008 presidential election allowed the KMT-nominated MYJ to achieve the ultimate political power. Democratisation was still the core geopolitical dimension of the state's grand strategy between 2000 and 2008, although CSB anchored the discourse of Taiwanese nationalism within Taiwanese democracy. It can be argued that the democratisation of the ROC was shown to have succeeded after 20 May 2008 when CSB peacefully transferred political power to MYJ, as democratisation passed the two-political-power-turnover test (Huntington 1993).

In other words, 'a democracy can be considered consolidated if it can survive two changes in ruling parties through post-transition elections' (Fell 2018: 35). It is also clear, with the benefit of hindsight, that the state under MYJ's governance complied with the ROC's constitutional settings. Thus, in modern Taiwan, democracy had already become 'the only game in town and when no-one can imagine acting outside the democratic institutions' (Przeworski 1991: 26, cited in Fell 2018: 35).

In clear contrast to CSB's *Fenghuo Waijiao* (Scorched-earth Diplomacy), which meant that the state governing Taiwan and supporting the TIM would openly strive to expand its national living space and publicly seek to increase the number of countries that granted it diplomatic recognition, MYJ emphasised the importance of his *Huolu Waijiao* (Viable Diplomacy) approach. '"Viable Diplomacy" involves two major initiatives—the "Diplomatic Truce" (*Waijiao Xioubing*) and "proactive diplomacy"' (Huang 2016: 468).

A foundation of MYJ's Diplomatic Truce was his declared prescription for alleviating the tensions of the cross-Strait relationship: *Bu Tong, Bu Du, Bu Wu* (No Unification, No Independence, No Use of Military Means; Su 2009). The major ingredient of the prescription was Ma's acceptance of the

One-China Principle, which was the bottom line set by the PRC for political parties and people conducting cross-Strait interactions and exchanges. Endorsing the One-China Principle, MYJ implemented his Diplomatic Truce to cease the political struggles against the PRC for international recognition of the ROC and to initiate a virtuous cycle for improving relations across the Taiwan Strait. MYJ also hoped to terminate the policy of chequebook diplomacy (Hsiao 2018; Huang 2016: 468).

The termination of chequebook diplomacy paved the path for MYJ's Proactive Diplomacy. Chequebook diplomacy as a metaphor, in brief, simply meant that the ROC vied for international recognition during the presidencies of LDH and CSB through offering clandestine subsidies to existing or potential allies. MYJ's endeavours to curtail this type of practice were based on his three principles for offering foreign aid: legitimate goals, legal procedures and effective implementation. During his administration, the ROC provided foreign aid not in exchange for international recognition but for necessary functional connections with and participation in international society, for example, the promotion of cultural exchanges, economic cooperation and participation in the World Health Organization (Hsiao 2018; Huang 2016: 471).

The general secretary of the CCP, HU Jintao, acknowledged MYJ's proposal for the truce. Upholding the One-China Principle, despite facing minor opposition when it attempted to participate in certain international activities, MYJ's Proactive Diplomacy was, by and large, not obstructed by the PRC. The CCP's public acknowledgement of MYJ's Diplomatic Truce in August 2008 and practical acquiescence afterwards were sustained by the CCP's trust in MYJ's continuous corroboration and support of the One-China Principle (Hsiao 2018: 165–166; Huang 2016); the trust was also partially based on MYJ's actions to create distance between the state and the TIM.

In addition, MYJ took action to demonstrate that his endorsement of the One-China Principle was not mere lip service. MYJ's so-called policy of de-Taiwanisation 'began when his inaugural address referred to people on both sides of the Strait as the *Zhonghua minzu* [Chinese nation], a term that implies Chineseness, defined by racial and cultural unity' (Hughes 2016: 164). He 'also appealed to Confucian values, making Classical Chinese and the Confucian canon required material for school courses again' (Hughes 2016: 164). Other minor but symbolic gestures included changing the name of the nationalised postal service from Taiwan Post back to China Post and reverting the National Taiwan Democracy Memorial Hall back to its original name, the Chiang Kai-shek Memorial Hall (Hughes 2016).[5]

Initially, the ideological focus of the ROC's grand strategy during MYJ's presidency was to downplay the role of ideology to concentrate the people's energy on economic prosperity and avoid stirring up any further resistance to KMT rule and economic cooperation with the Chinese mainland. 'After 2005, the KMT under its new chairman Ma Ying-jeou sought to freeze the identity battle and shift popular attention to the economic performance of the DPP government' (Wu 2016: 403), which contributed to an unprecedented economic downturn, as shown earlier. He succeeded in mobilising 'the economy-concerned, identity mobilization-fatigued population', persuading them with the notion that it was necessary 'to save Taiwan's economy by improving relations with the mainland' (Wu 2016: 403).

Eventually, MYJ's grand strategy failed not only to prevent the provocation of anti-China sentiment but also to gather sufficient political support for his efforts to globalise Taiwan's economy through China (Leng 2016). Shortly after MYJ was elected president, 'the majority began to identify themselves as "Taiwanese only" for the first time. By the time he ran for re-election in March 2012, this figure had reached a new high of 54%' (Hughes 2016: 165). His ideological passivity and active promotion of economic cooperation with the Chinese mainland generated anxiety and fear in Taiwan's population about the CCP's gradual control of the vitality of Taiwan's economy and political autonomy. This apprehension and terror fuelled the rise of Taiwanese national identity and 'a common rejection of the idea that Taiwan is a part of China' (Hughes 2016: 165). The worry and phobia reached their apex in the so-called Sunflower Movement between 18 March and 10 April 2014 (Jacobs 2016: 65).

Effectively, 'globalization through China' (or the globalisation of Taiwan's economy through the Chinese mainland market) encapsulated the economic aspect of the ROC's grand strategy during MYJ's presidency (Leng 2016: 235). His ideologically passive, economically oriented grand strategy, however severe the anti-China sentiment it provoked, was seen as an attempt to adjust Taiwan's economy and stimulate its stagnating growth via formalised economic cooperation with the Chinese mainland (Leng 2016).

The rationale behind MYJ's formalisation of cross-Strait economic cooperation was the real reason for the economic transactions between the Chinese mainland and Taiwan. Before 2001, when CSB finally abandoned the official prohibition against investments by Taiwan's capitalists, those capitalists had invested on the mainland through Hong Kong or through shell companies registered in third countries such as the British Virgin Islands or the British Cayman Island, to evade the ROC's legal restrictions against investments on the mainland. Nonetheless, Taiwan's capitalists formed a production chain with the US and Japan on the one hand, and with the Chinese mainland, on the other (Lee and Chu 2016). They were able to connect American or Japanese technologies, Taiwan's know-how, and the production capacity of the Chinese mainland (Leng 2016). Furthermore, Taiwan's capitalists who invested in the Chinese mainland confronted no cultural or linguistic barriers that could attract 'Japanese as well as U.S. firms to seek strategic alliances with them to carry out investment in China' (Lee and Chu 2016: 413).

Under such a circumstance, MYJ promoted the Three Links policy to implement direct postal, trade and transportation connections between Taiwan and the Chinese mainland. The Three Links policy was viewed not only as a springboard for developing mutual trust across the Strait but also as an initial approach to the establishment of new transportation routes with other major countries. For example, new aeroplane transportation routes with other East Asian countries were established. These routes were expected to make international trade with the Chinese mainland more convenient through Taiwan's newly acquired connections (Leng 2016).

MYJ promoted the establishment of the Economic Cooperation Framework Agreement (ECFA) with the PRC, which was a policy that had been promised during his presidential campaign. When the ECFA was signed on 29 June 2010, it can be argued that the 'new China policy turned market-driven integration into government-promoted integration. The ECFA led to [formalised] contacts' (Lee and Chu 2016: 420). The optimistic rationale behind the ECFA was to use it 'as a platform to attract foreign companies to explore the Chinese market collectively' (Leng 2016: 244). MYJ's administration expected that the ECFA would function as well as the Three Links policy in terms of facilitating the globalisation of Taiwan's economy through the market on the Chinese mainland. 'The optimistic viewpoint argues that the ECFA could serve as a catalyst to Taiwan's integration with regional economic structures' (Leng 2016: 244).

However, the formalised enhancement of cross-Strait economic cooperation did not produce momentum for further political integration and identity unification between the Chinese mainland and Taiwan. The DPP in Taiwan worried that MYJ's policy of globalisation through the mainland market would have increased Taiwan's economic vulnerability to the PRC's malicious intentions. The DPP complained that 'Ma showed no signs of balancing growing economic dependence on the PRC with the building of Taiwanese consciousness and democratic institutions' (Hughes 2016: 165).

Additionally, the DPP claimed that MYJ's approach to economic cooperation with the Chinese mainland would have caused more job losses as a result of investments by mainland Chinese capitalists. The DPP also claimed that MYJ's promotion of globalising Taiwan's economy through the mainland market would have privileged only a few, 'while the interests of the "many" [would not have been] represented in the Taiwan-China nexus' (Lee and Chu 2016: 421).

Finally, the DPP disseminated the idea that economic cooperation with the Chinese mainland would inevitably lead to the annexation of Taiwan by the PRC (Lee and Chu 2016). This propaganda was highly effective because the PRC leaders did not publicly deny such a connection. Thus, the so-called Sunflower Movement erupted and was supported by the DPP. MYJ's plan to sign the Cross-Strait Agreement on Trade in Services and the Cross-Strait Agreement on Trade in Goods was blocked in April 2014 (Lee and Chu 2016: 420).

Nevertheless, the cross-Strait relationship reached its most stable and friendly phase during MYJ's presidency. The amicable atmosphere between the two governments across the Taiwan Strait was evidenced at the meeting between the general secretary of the CCP, XI Jinping and MYJ in Singapore on 7 November 2015 (Hsiao 2018: 342). During this meeting, the military forces were employed differently from the time when LDH and CSB governed; consequently, the military's construction of the ocean also changed, which is the subject of the next subsection.

The Construction of the Ocean by the ROC's Military Forces between 1991 and 2016

The alteration of the ROC's military strategy in the 1990s was closely related to the high politics at the centre of the state's grand strategy. *Fangwei Guoshou, Youxiao Hezu* (Resolute Defence, Effective Deterrence), was officially presented in 1995. The new military aspect of the ROC's grand strategy 'represented not only a response to differing external and environmental circumstances, but also a different stage in the evolution of a continuous struggle for dominant strategic/operational concepts among and between the three ROC Armed Services' (Edmonds and Chen 2003: 91).

The interservice rivalry was but one reason for the change. The ROC implemented its largest military modernisation project during the 1980s and 1990s, named *Guo Jun Er Dai Bingli Zhengjian* (the Construction of the Second-Generation Armed Forces) and provided the ROCA and ROCAF more budget than the ROCN, 'raising these two branches to higher levels than the Navy' (Yang 2003b: 89). However, the ROCN did not come out a total loser in these manoeuvrings. The ROCN purchased two submarines from the Netherland in 1983 and was also provided with funds to build Perry-class guided-missile frigates in Taiwan with the technological assistance of the US after 1987, although the actual building did not begin until the early 1990s. S-70C anti-submarine helicopters and S-2T aircraft were bought from the US in response to the PRC submarines. More importantly, *Hsiung Feng* II all-weather anti-ship missiles with dual-homing capability through active radar and imaging were developed in the late 1980s and produced in the early 1990s (Yang 2003b: 87–88). Therefore, it can be argued that the interservice rivalry was more related to the strategic incompatibility of the ROCA general HAO Bocun and ROCN admiral LIU Heqian, which could lead to mutually inconsistent pathways of military development.

The change of the military aspect of the grand strategy was more related to the politics at the apex of the state's centre. When LIU Heqian was appointed as the chief of the general staff in 1991 by LDH, HAO Bocun had already been promoted to the premiership. LDH used Liu to diminish Hao's influence over the military, as suggested earlier. Furthermore, as was also discussed previously, HAO Bocun's military focus was defensive, concentrated on the main island of Taiwan. He aimed to annihilate the PLA invasion forces on the shores of Taiwan, for which the maintenance of the ROCA's air supremacy was crucial. The ROCA and ROCAF played vital roles in the defensive military strategy.

On the contrary, LIU Heqian did not agree with the secondary role of the ROCN in Hao's approach. Instead of being deployed to intercept the

PLA's landing and logistical transports, Liu believed that the ROC's military should use the Taiwan Strait as the first line of defence against the PLA, highlighting air supremacy and command of the Taiwan Strait. The ROCN and ROCAF were expected to deal with the military blockade by the PLA rather than a direct, massive scale of amphibious assault in Liu's plan. Furthermore, Liu intended to use the naval force to protect the SLOCs and convoy homebound fleets from South China to Taiwan.[6] More critically, due to the geographic conditions of Taiwan Island, the scenario of a large-scale amphibious attack against the ROC could only take place along the western coast of Taiwan, which, incidentally, was the most developed area on which the whole economy of Taiwan depended. In Liu and his followers' view, a final military showdown on Taiwan, even along the coast, would devastate the whole island (Edmonds and Chen 2003).

Notably, in December 1994, ROCA general (and long-time political rival of HAO Bocun) JIANG Zhongling was appointed as the minister of national defense. In July 1995, ROCA general LUO Benli was appointed as the chief of the general staff. Jiang and Luo readjusted the ROC's military forces amidst the hostile cross-Strait relationship in July 1996 according to the principle: 'Resolute Defence, Effective Deterrence' (Huang 2017: 45–46, 104). Such a readjustment could be viewed as a response to the 1995–1996 Taiwan Strait Crisis.

In July 1995, the PLA launched missiles aiming at the sea area approximately 60 kilometres northeast from the northern Taiwan port city, Keelung, and began a series of large-scale military manoeuvres. Missile tests and military manoeuvres were conducted again in March 1996. The ROC's military forces realised their strategic vulnerability to ballistic missile attacks and proposed the acquisition of larger, Aegis-equipped US destroyers as an extra layer in the anti-tactical ballistic missile system defending Taiwan. The ROCN also requested the purchase of more naval vessels, including submarines, to expand its fleets (Edmunds and Chen 2003: 103; Huang 2017: 98).

It is worth noting that such a military strategy corresponded well with LDH's connection of the state with the TIM. In the antagonised cross-Strait relationship and under the rule of a state leader, LDH, pursuing the goals of the TIM and attempting to cooperate with and co-opt the TIM forces in an era of democratisation, the ROC's military increased the volume of its offensive tone. Having acquired more advanced weapons to improve the quality of its weaponry in line with its plan to construct the second-generation military (Yang 2003a: 60), the ROC's armed forces became confident in its capacity of conducting military retaliation and counter-strikes. Thus, in 1995, the military aspect of the grand strategy changed from one of staunch defence to 'Resolute Defense, Effective Deterrence'. The term 'deterrence' in such a context contains an aggressive, hostile connotation in the Chinese language. Adding such a term to its strategic guideline reflected its faith that the 'ROC Armed Forces developed [their] basic combat capabilities [and] skills in performing effective defensive operations in 1995', as the Ministry of National Defense (2006: 99) suggests.

With such a strategic guideline gradually shifting towards an aggressive posture in the name of deterrence, the ROCN was charged with securing the command of the sea of the Taiwan Strait, around the marine areas of Taiwan Island and along the SLOCs between Taiwan and the South China Sea. The ROCN 'was committed to introducing new missiles, automating its command and control systems and modernizing its anti-submarine warfare capability' (Yang 2003a: 62).

Critically, a reason for the ROCN securing the command of the sea over the Taiwan Strait was to form a thick defensive layer protecting Taiwan Island from the PLA's assaults. The depth of defence then increased to cover the whole Taiwan Strait. The ROCN was expected to prevail in naval battles and anti-blockade missions (Tsai 2003). Thus, the ROC's military forces began to gradually convert their construction of the ocean after 1991 into the style of the original Mediterranean Sea Model. The construction was closely related to the state's grand strategy and the changing balance of military power between the two sides of the Taiwan Strait.

By the time CSB became the ROC president in May 2000, the cross-Strait relationship had continued to deteriorate and had become more antagonistic. The relationship reached its nadir when CSB pronounced his 'one country on each side of the Taiwan Strait' rhetoric *cum* policy in August 2002. However, in terms of the preparation for war against the Chinese Communists and PLA, he announced his military strategy *qua* conviction *Juezhan Jing Wai* (Seeking Decisive Battle outside the Territory) as soon as the inauguration of his first presidential term took place (Edmonds and Chen 2003: 100). In the same year, the Ministry of National Defense (2000: 27) presented the guiding military principle of the ROC's grand strategy: *Youxiao Hezu, Fangwei Gushou* (Effective Deterrence, Resolute Defence). CSB's military strategy had a stronger, more aggressive emphasis than that during the LDH presidency.

The *National Defense Reports* of the years 2000 and 2004 both concluded that the PLA had already started military modernisation and that it would be transformed into an offence-oriented military power with a strong green-water naval force. They also suggest that the PLA was inclined to attack Taiwan via the following approaches: coercive occupations of Taiwan's offshore islands, shock intensive missile attacks and a long-term blockade by naval and air forces. Notably, the *National Defense Reports* published during CSB's presidency all acknowledged the possibility of the PRC using military means to coerce the people in Taiwan into surrender and acceptance of its terms for reunification. However, none of the reports mentions the chance of a direct, large-scale PLA amphibious assault and landing for permanent occupation. During this period of time, the ROC's military forces 'believed that [the PRC lacked] the ability to invade and successfully occupy Taiwan and [was] unlikely to gain that capability in the foreseeable future' (Stocker 2003: 141). Additionally, these reports seemed to imply that the ROC Armed Forces believed that the US, and possibly Japan, would intervene militarily.

Under the guidance of CSB's 'Seeking Decisive Battles outside the Territory', effective deterrence became much more important than resolute defence. Thus, the military strategy followed the rationale of the second-strike (with conventional rather than nuclear weapons) to achieve its goal of effective deterrence, and 'the roles of the ROCAF and the ROCN were given particular emphasis' (Edmonds and Chen 2003: 100). In September 2004, then ROC premier YOU Xikun even proclaimed his view that the ROC's military would use missiles to attack Shanghai and other assets on the Chinese mainland once the PLA took military actions against Taiwan (Bush 2016; Huang 2019). If the PRC had deployed military means against Taiwan, ROC military forces under CSB's guideline would have gone 'on the offensive, at least tactically in support of strategic defense' (Bush 2016: 39), that is, attacking civilian as well as military targets on the Chinese mainland with ballistic and cruise missiles, and air and naval forces.

CSB's regime also pursued the goals of the TIM, closely cooperating with and successfully co-opting the societal movement. Adopting a hostile grand strategy and military strategy against the PRC was a natural consequence and beneficial for mobilising more mass support for its regime. Having equipped itself with more advanced weapons, such as Kidd-class destroyers, *La Fayatte*-class frigates, Knox-class frigates, domestically built Perry-class frigates and other missile and anti-submarine systems (Grove 2003), the ROCN seemed to be confident in assuming military responsibility of the grand strategy of CSB's regime.

In such a strategy, the ROCN not only had to secure the command of the sea under the assistance of superior air power but also, with the arrival of Kidd-class destroyers between 2005 and 2006, became 'the air battle command/control unit in guiding the ROCAF's aircraft in over-the-horizon attack and thus extend the ROCAF's capability for force projection' (Edmonds and Chen 2003: 104). Additionally, the ROCN had to be able to launch missile and amphibious attacks against the enemy targets on the Chinese mainland as a critical pillar supporting CSB's aggressive deterrence strategy. Therefore, having evaluated its role in the strategy, it can be argued that the ROCN in particular and the ROC's military forces in general between 2000 and 2008 constructed the ocean in accordance with the original Mediterranean Sea Model.

Being confident in their ability to attack the Chinese mainland, however, the ROC Armed Forces' chance of actually achieving such a strategic objective continued to dwindle because the PLA was constantly strengthening its military capacity. It is worth a direct quote from a professor of the US Naval War College, who wrote in early 2008:

> China's recent military modernization has fundamentally altered Taiwan's security options. New Chinese submarines, advanced surface-to-air missiles, and, especially, short-range ballistic and land-attack cruise missiles have greatly reduced Taiwan's geographic advantage. Taipei can

> no longer expect to counter Chinese military strengths in a symmetrical manner, with Patriot interceptors, diesel submarines, surface warships, F-16 fighters, and P-3 maritime patrol aircraft. Taiwan must therefore rethink and redesign its defense strategy, emphasizing the asymmetrical advantage of being the defender, seeking to deny the People's Republic its strategic objectives rather than attempting to destroy its weapons systems.
>
> (Murray 2008: 2)

Additionally, according to the assessment by a US Department of Defense annual report, the PLA still lacked adequate preparations and optimal capacity to launch a full-scale amphibious invasion of Taiwan Island (Bush 2016). Therefore, it convincingly suggested that

> the only defense strategy that makes sense for Taiwan [was] one that better exploits its key strategic feature: the fact that it is an island; and one that exploits one of the PLA's greatest weakness, that is, the vulnerability of an invasion force to attack while it is transiting the Taiwan Strait and coming ashore on Taiwan.
>
> (Bush 2016: 540)

Following such a rationale, the porcupine strategy was advised. As William Murray (2008) clearly points out, the porcupine strategy for the ROC's defence of Taiwan required military-civilian cooperation to stockpile critical energy, food and medical supplies and to establish rationing institutions to prepare for a possible PLA blockade. Murray (2008) also convincingly suggests that the ROCAF and ROCN could hardly survive under the PLA's saturation attacks with its new ballistic and cruise missile systems. Therefore, the porcupine strategy relied heavily on the army as the ultimate military force to repel the PLA's amphibious invasions.

Murray (2008) identifies several points that the ROC's military could follow. First, the ROCA should be well trained and motivated. Second, an all-volunteer army could strengthen the security of Taiwan, for it could better utilise the weapons of high-end technology. Third, the ROCA should also acquire more mobile missiles, multiple launch systems and attack helicopters as well as advanced tanks, artillery and anti-tank weapons. Fourth, surf-zone sea mines should be deployed in due course. Fifth, the ROC and its military forces should fortify the important civil and military facilities, build redundancy and decoys for crucial infrastructure and make critical weapon systems mobile to increase their chance of survival.

Although Murray's (2008) suggestions were not fully followed by the ROC's armed forces, the administration of the then newly elected MYJ shifted the military aspect of the state's grand strategy from CSB's 'Effective Deterrence, Resolute Defence' to *Fangwei Gushou, Youxiao Hezu* (Resolute Defence, Credible Deterrence; Ministry of National Defense 2011: 20).

Moreover, MYJ provided a guideline for the ROC Armed Forces: *Gu ruo Panshi* (Being as Solid as Bedrock; or the official English translation: 'Hard ROC'; Ministry of National Defense 2009: 6). CSB's approach of 'Seeking Decisive Battles outside the Territory' was abandoned.

This military dimension of the ROC's grand strategy of MYJ's presidency was consistent with the other strategic aspects. As elaborated earlier, MYJ aimed to diminish the influence of the ideology of the TIM, focus society's energy on economic development and develop a good rapport with the Chinese Communists. His adherence of the One-China Principle and adoption of the Diplomatic Truce policy stabilised the cross-Strait relationship to a large extent. Therefore, it was sensible to formulate a corresponding passive defensive policy.

During MYJ's presidency, the ROC not only endeavoured to foster friendly cross-Strait relations, but also adopted 'a more practical position that is best described as "hedging"' (Wu 2018: 547). Hedging refers to a tactic of smaller states to reduce their security risks from leaning too much towards a certain stronger state by establishing an amicable relationship with another stronger state in an attempt to balance the influence from the former. MYJ's hedging position *qua* policy was implemented by purchasing US weapons, especially during his second term of presidency. The amount of money spent on arms acquisition from the US in millions reached US$425, US$553, US$1081 and US$638 million in the years 2012–2013, 2013–2014, 2014–2015 and 2015–2016, respectively. The list of items bought from the US included Patriot Advanced Capability III missiles, Hellfire anti-armour missiles, Apache helicopters, Black Hawke helicopters and Phalanx close-in weapon systems for the domestically built Tuochiang-class corvettes. These weapons arrived in Taiwan in 2014. This set of purchases could balance the ROC-US and the Chinese mainland-Taiwan relationships (Wu 2018: 528). Notably, the items obtained from the US were close to what Murray (2008) suggests.

Guided by MYJ's grand strategy, the military strategy readjusted and requested every branch of the ROC's military to conduct joint defence operations. The joint operations, which coordinated through the joint command, control, communications, computers, intelligence, surveillance and reconnaissance system, aimed to utilise asymmetrical and innovative warfare tactics and techniques to, first, strike the enemy forces from a long distance and, eventually, destroy them on the shore of Taiwan Island. Local air supremacy, sea control and ground domination were to be jointly secured (Ministry of National Defense 2009: 10–11).

Similar to the military strategy designed by HAO Bocun, the final military showdown would occur on the beaches of Taiwan. Different from Hao's design were the roles of the ROCN and ROCAF, at least on paper. The new roles of the ROCN and ROCAF involved taking a critical part in the joint operations to repel the invasion along the coastal areas of Taiwan. Because of the immense improvement of the PRC's military technology and power, Taiwan's resources could not sustain a symmetrical arms race against the

Chinese mainland. Because of MYJ's efforts and intention to ameliorate the tension in the cross-Strait relationship, it was sensible for the ROC's military to fortify its deployment of missiles and adopt a passive defensive posture, which could make the cost of a potential PLA amphibious invasion of Taiwan Island too exorbitant for the PRC to afford.

In such a military strategy, the ROCN was expected to develop into a force with the capacities of 'stealth, speed, long-range, mobility, precision, intense fire power, and superior protection' (Ministry of National Defense 2013: 46). Such an abstract proclamation can be exemplified by the establishment of a fleet of *Kuang Hua* VI-class missile boats and the development of the *Tuochiang*-class corvettes in 2010 and 2015, respectively. Both marine and terrestrial coastal areas were the expected locations; when the PLA crossed the Taiwan Strait and reached the coastal areas was the time that the ROC Armed Forces picked to initiate their counterstrike.

Therefore, the romance of sea battles conducted by the ROCN's flagships as centres of extraterritorial strike forces assisted by the ROCAF's fighters was renounced. In contrast, the defensive functions of the ROCN and ROCAF were strengthened by their integration into the military strategy of joint operations together with the ROCA, exploiting the advantages of Taiwan's geography (that is, the sea of the Taiwan Strait) and the most serious weakness of the PLA during transport. Thus, in the light of the Steinbergist Classification, the ROC's military, highlighting the importance of missiles, anti-missiles and joint operations with a passive defensive strategy, changed their construction of the ocean back to the Indian Ocean Model.

A Brief Summary for the Final Comparison

I can now summarise this section through the double lenses of the ROC's grand strategy and the changing balance of military strength across the Taiwan Strait between 1991 and 2016. From 1991 (or more precisely, 1988) to 2000 when the ROC was ruled by LDH, the consolidation of presidential power was linked with the democratisation of the state and, later, with the TIM. LDH exacerbated the cross-Strait relationship in the latter half of the 1990s, pushing for economic globalisation without the market on the Chinese mainland. When CSB gained presidential power, by and large, he continued LDH's path to economic globalisation without China. Moreover, he maintained the connection of the state with the TIM and adopted a hostile approach towards the PRC.

During LDH's and CSB's rule, the ROC's military forces shifted their terrestrial-centric strategy against amphibious landing to a maritime-centric strategy deploying naval flagships as pivotal units coordinating with air and other naval forces to undertake military retaliation from the Taiwan Strait against targets on the Chinese mainland. The ROCN believed that the PLA's missile technology and capacity would nullify any army defence on the land of Taiwan Island. However, during this period, with the receipt of new weapons, the ROCN was confident that it could defeat the PLA at sea with the assistance of the ROCAF. Thus, in the view of the ROCN, it was sensible to use the Strait as an arena in which decisive battles and counterattacks could be conducted. The ROC's military constructed the ocean consistently with the original Mediterranean Sea Model.

In contrast, when MYJ became president, he endeavoured to separate the state from the TIM. He made efforts to downplay the role of ideology and to concentrate the energy of the society of Taiwan on economic development and globalisation through the Chinese mainland market. MYJ also strove to make his Diplomatic Truce with the CCP successful. Observing the One-China Principle, the cross-Strait relationship reached its best condition since 1949. This largely reduced the danger of war between the two sides of the Strait and allowed the ROC Armed Forces to pursue a passive defensive strategy.

Guided by the principle of 'Being as Solid as Bedrock', the ROC's military emphasised the importance of joint operations of its three branches; the strengthening of the PLA and improvements in its military technology were such that the ROCN and ROCAF would likely be overwhelmed both on and above the Taiwan Strait. Thus, the ROC's military during the MYJ presidency turned its focus to the shores of Taiwan to exploit the weakness of the PLA's maritime transport and amphibious landing capability. The ROC Armed Forces' joint operations against such a weakness was the key to its asymmetrical defence. In the view of the ROC's military forces, the Taiwan Strait became a source of danger as well as a space barricading against the enemy, and their construction of the ocean could, thus, be classified as the Indian Ocean Model.

Conclusion

In this chapter, I have used the grand strategy and military strategy concepts to examine the history of post-war Taiwan. The four important aspects of the ROC's grand strategy in the three phases (from 1949 to 1971, 1971 to 1991 and 1991 to 2016) have been elaborated. Such an historical narrative has singled out two critical structural factors influencing the ROC Armed Forces' constructions of the ocean after it retreated to Taiwan, Kinmen and Matsu in 1949. These two critical structural factors are the state's grand strategy and the military balance between the two sides of the Taiwan Strait.

This chapter, therefore, can contribute to Steinberg's (2001) analysis of the modern navies' construction of the ocean. This chapter suggests that a modern naval force does not inevitably construct the ocean in the fashion of the original Mediterranean Sea Model to justify its existence and share of the state budget. Instead, as illustrated by the case of the ROCN, such a naval force can construct the ocean under the heavy influence of the state's grand strategy and the balance of power between its military forces and those of the enemy.

As elaborated earlier, the overarching approach of a state to achieve its ultimate goals of national security is defined as the state's grand strategy. The military strategy as an important concept of geopolitics is usually seen as one rank below the grand strategy and is generally developed to obtain the grand strategic goals. In terms of the grand strategy, JJS, as an ardent supporter of the TPP claiming to reunify China under this official ideology, strove to build a strong economic foundation on Taiwan to sustain his military pursuit of removing the Chinese Communists from power on the Chinese mainland. JJS planned to initiate this mission after he consolidated his internal political position and external political support, mainly of the US.

Under the command of JJS, the ROC Armed Forces prepared themselves for an all-out military attack against the PLA. The ROC aspired to restore its rule on the Chinese mainland by removing the CCP. The ROC's military forces also undertook many small-scale amphibious assaults and military infiltrations while JJS was waiting for the most favourable occasion to launch the strike. The ROCN was then responsible for ensuring the command of the sea, the security of transportation of military materials and personnel, and a successful amphibious landing when the full-scale invasion began. This was the situation before 1965, when the ROC's armed forces constructed the ocean in accordance with the original Mediterranean Sea Model.

After 1965, JJS realised that the power of the PLAN had been greatly fortified and that the ROCN could no longer claim command of the sea in the Taiwan Strait or along the coastal areas of the Chinese mainland. JJS also understood that the plan to restore the ROC's governance on the mainland could not succeed without US support. The ROC's military ceased its operations against the PLA after 1965 and gradually changed its construction of the ocean towards the Indian Ocean Model. With hindsight, the hope of recovering the Chinese mainland for the ROC completely disappeared in October 1971 when the ROC was forced to leave the UN and the PRC was recognised as the solely legal representative of the entire Chinese nation.

The grand strategy of the ROC became offensive in terms of its political and ideological pursuits but defensive in terms of its military endeavours. The ROC still proclaimed that the TPP was the only legitimate doctrine according to which the whole Chinese nation should be governed. During JJG's rule, the state continued to strengthen the economy in Taiwan and endeavoured to create an industrialised capitalist system that could sustain rapid growth.

Politically, the JJG gradually loosened the state's control over Taiwan's society. The incremental liberalisation of society, together with JJG's plan to transition to a democratic regime, also served ideological functions. JJG hoped that the people of Taiwan would not join the TIM and would instead support the ROC and its mission to rule the Chinese mainland in accordance with the TPP. JJG also hoped to demonstrate the superiority of the ROC's governance to the people on the mainland and the CCP.

Consistently, the ROC's military switched its position to the defence of the main island of Taiwan. Anxious about being annihilated by the PLA's saturation attack on and above the sea of the Taiwan Strait, the ROC's armed forces relied on the ROCAF and ROCA and aimed to destroy the enemy on the shores of Taiwan. The ROCN's role became auxiliary. The military thus constructed the sea in the style of the Indian Ocean Model between 1971 and 1991.

Nevertheless, the ROC's grand strategy changed after JJG died in 1988. In the late 1980s and early 1990s, the grand strategy of LDH was to consolidate his presidential power. The societal forces demanding democratisation and the TIM were co-opted by LDH. Since the mid-1990s, he had publicly incorporated and created linkages between the state and TIM; this trend continued and was even deepened by CSB after 2000. Both LDH and CSB attempted to promote Taiwan's economic globalisation without utilising the market on the Chinese mainland, fearing that they would increase the vulnerability of Taiwan's economy and diminish the opportunity to establish an independent sovereign state under the name Taiwan. Moreover, their promotion of the TIM had already severely antagonised the cross-Strait relationship.

The ROC's military under the command of LDH and CSB also began to adopt an aggressive stance poised to implement military retaliation against both civil and military targets on the Chinese mainland in an attempt to deter PLA military operations. Encouraged by the newly acquired weapons and confident in their ability to defeat the PLAN, the ROCN was confident in its central role in sea battles. The Taiwan Strait, rather the shores of Taiwan Island, became the front line in the military strategies during the reigns of LDH and CSB. The ROC's military forces thus constructed the ocean consistently with the original Mediterranean Sea Model.

The optimistic view of the military forces led by LDH and CSB vanished after 2008 when it became clear that Taiwan could not sustain an arms race against the Chinese mainland and that the PLA's missile, naval and air forces had acquired the capacity to defeat the ROCN-led and ROCAF-assisted forces in the Taiwan Strait, as the US defence report indicated. Furthermore, Taiwan's economy deteriorated under CSB's rule. MYJ thus made efforts to dilute the ideological effects of the TIM in Taiwan's society and divert the energy of the population to Taiwan's economic development and globalisation through the mainland market.

Additionally, not only did MYJ continued to purchase military weapons from the US, but also disconnected the state from the TIM and adhered to the One-China Principle to improve the cross-Strait relationship. Fostering an amicable atmosphere for the interactions between the two sides of the Strait was expected to minimise the possibility of the PLA imposing a military blockade against Taiwan. It was also intended to maximise the viability of the 'Hard ROC' strategy, which aimed to pull the location of ultimate, decisive battles back to the coastal areas of Taiwan, relying on joint military operations of the three branches to defeat the PLA's amphibious invasion. Therefore, the ROC Armed Forces commanded by MYJ constructed the ocean in line with the Indian Ocean Model.

The above summary of this chapter supports two comparisons. The first comparison is amongst the three phases: from 1949 to 1971, from 1971 to 1991 and from 1991 to 2008. The second comparison is amongst another three phases: from 1991 to 2000, from 2000 to 2008 and from 2008 to 2016. Two critical structural factors (the state's grand strategy and the balance of military power between the two sides of the Taiwan Strait) are the focal points of these two comparisons, and the purpose of the comparisons is to demonstrate an important thesis of Steinberg's (2001) theory of territorial political economy. That is, the social constructions of the ocean can be influenced by the geographical conditions of the ocean and the material organisation of society.

In the phases (as cases for comparisons) of this chapter, the Taiwan Strait is notoriously difficult for military forces to cross during wartime. The places suited for an amphibious landing on the main island of Taiwan are few and concentrated on the west side of the island. These are the given geographical conditions. In addition, between 1949 and 2016, the PLA's amphibious landing and troop transport capacity remained relatively insufficient to accomplish its mission of a permanent occupation of Taiwan. This was a given condition of the material organisation of society on the Chinese mainland.

In the first comparison, and in the first case between 1949 and 1971, the ROC's grand strategy was generally inclined to invade the Chinese mainland and its naval power was stronger than that of the PLAN. Additionally, in the second case between 1991 and 2008, the ROCN was confident in its ability to defeat the PLAN at sea and the ROC adopted an aggressive posture to deter the PRC and PLA. In contrast, the grand strategy and the military forces of the ROC between 1971 and 1991 became staunchly defensive. The ROC could no longer pursue its goal of restoring its rule over the mainland militarily and decided to take advantage of the weakness of the PLA and the defensive utility of the Taiwan Strait. Thus, it is clear that when a state adopts an aggressive military posture and enjoys relatively stronger naval *qua* military power, the construction of the ocean by the armed forces tends to be aligned with the original Mediterranean Sea Model. When the state's naval *qua* military power is relatively stronger and the grand strategy is defensive using the sea as a spatial barrier against its opponent, the ocean can be constructed in the style of Indian Ocean Model.

In the second comparison, the ROC's grand strategies in the two phases, from 1991 to 2000 and from 2000 to 2008, were aggressive against the Chinese mainland, albeit limited to second strikes and military retaliation rather than offensive invasions. Moreover, the ROC military, especially the ROCN, was confident about its power. They received weapons and completed many military modernisation and arms development projects during the period of JJG's governance. The ROC's naval and air forces aimed to prevail in decisive battles at sea in the Taiwan Strait. In contrast, MYJ's administration strove to ameliorate the relationship between Taiwan and the Chinese mainland. During this period, the ROC's grand strategy towards the CCP was friendly. The ROC's military did not wish to undertake highly risky sea battles, having realised that they could not win a symmetrical confrontation with the PLA. Instead, they prepared for battles on the shores of Taiwan. The ROCN would participate in joint operations against any amphibious enemy invasions. Therefore, similar to the result of the above first comparison, when a state's grand strategies are aggressive and the balance of military power favours the state, its military tends to construct the ocean in the manner of the original Mediterranean Sea Model. When a state's grand strategy becomes defensive and the balance of military power favours its opponent, a state's armed forces construct the ocean in accordance with the Indian Ocean Model.

In the above comparisons, the two structural factors can be categorised as what Steinberg (2001) calls the material organisation of society, and the Taiwan Strait became a geographical condition. Thus, not only have the above comparisons (and effectively this entire chapter) modified and amplified Steinberg's (2001) simplified critique of the construction of the ocean by modern navies, but they also have illustrated an important thesis of his theory of territorial political economy. That is, the social construction of the ocean can be influenced by the geographical conditions of the ocean and the material organisation of society that utilises, regulates and represents the ocean. Indeed, Steinberg (2001) also explicates that the influence can flow in the opposite direction. This thesis is elaborated in the next chapter.

Notes

1 This chapter is a modified and greatly expanded version of my journal paper, see Hou (2020).

2 Chung (2012: 155) believes that the PLA's nuclear test that succeeded in October 1964 became a reason for JJS to alter his offensive military posture to a defensive one. The PLA acquired airborne nuclear bombardment ability after May 1965 and obtained nuclear missile delivery capacity after October 1966. However, JJS still attempted to implement his military infiltration plan via the sea in August 1965, which led to the debacle of the Dongshan Naval Battle. It is sensible to take the PLA's nuclear ability into account, but the loss of the command of the sea had already resulted in JJS changing his grand strategy before the PLA could successfully deliver its nuclear bombs and warheads.

3 Following Fell (2018), liberalisation is viewed differently than democratisation in this monograph. Liberalisation of society refers to 'some increase in freedom of speech and the press. [The incumbents] may allow opposition politicians to win office in election, though they do not relinquish ultimate control' (Jacobs 2008: 51, cited in Fell 2018: 40). According to Freedom House ratings in the 1970s, Taiwan society became partly free in 1977 (Fell 2018: 37). In the 1980s, the candidates challenging the KMT repeatedly won 'at local executive, provincial assembly and supplementary national level elections during this decade' (Fell 2018: 36). JJG also tolerated the establishment of the Democratic Progressive Party (DPP) in September 1986 (Fell 2018: 36).

4 The Two Trillion and Two Stars Program was 'a grandiose industrial project involving NTD 2 trillion-worth of investment in four strategic sectors: DRAM (dynamic random access memory), liquid crystal panels, LCD, solar power, and biotechnology' (Lee and Chu 2016: 415).

5 MYJ, nevertheless, apologises in his memoir for not adequately removing the contents of de-Sinicisation from school curricula during his presidency (Hsiao 2018: 239).

6 This point was very different from HAO Bocun's view. In Hao's (2019: 290) memoir, he clearly states that the strength of the ROCN would not be reduced by a counter-blockade. Lighter, faster fishing boats could be employed to conduct maritime guerrilla warfare and sabotage the SLOCs of the Chinese mainland.

References

Biddle, Stephen (2007) 'Strategy in War', *PS: Political Science and Politics*, 40(3), 461–466.

Bush, Richard (2016) 'Security Policy', in Gunter Schubert (ed.), *Routledge Handbook of Contemporary Taiwan* (Oxon; Routledge), 529–543.

Chang, Yu-fa (2001) *Zhonghua Minguo Shigao*, xiuding ban (A Draft of the History of the Republic of China, revisited edition) (Taipei; Linking).

Chi, Mengbin (2013) *CHI Mengbin Xiansheng Koushu Lishi* (The Oral History of Mr CHI Mengbin), recorded by Li Chang and Chin-lan Tseng (Beijing; Jiuzhou Press).

Chu, Wan-wen (2017) *Taiwan Zhanhou Jingji Fazhan de Yuanqi: Houjing Fazhan de Weihe yu Ruhe* (The Origin of Economic Development in Postwar Taiwan) (Taipei; Academia Sinica).

Chung, Chien (2003a) *Zai Jingtao Hailang Zhong Zhanbei Hangxing: Haijun Jiangting Zhi* (Sailing in the Stormy Sea with Combat Readiness: A History of Naval Vessels) (Taipei; Rye Field Publishing).

Chung, Chien (2003b) 'A Review of the ROC's Military Reform: A Case Study of the Taiwanese Navy', in Martin Edmonds and Michael M. Tsai (eds.), *Defending Taiwan: The Future Vision of Taiwan's Defence Policy and Military Strategy* (Oxon; Routledge), 114–126.

Chung, Chih-tung (2012) *The Evolution of Taiwan's Grand Strategy: From Chiang Kai-shek to Chen Shui-bian*, PhD Thesis (London; London School of Economics and Political Science).

Corbett, Julian (1911) *Some Principles of Maritime Strategy* (London; Longmans, Green and Co.).

Edmonds, Martin and York W. Chen (2003) 'An Assessment of the ROCN's Modernization Program: Strategic and Operational Considerations', in Martin Edmonds and Michael M. Tsai (eds.), *Taiwan's Maritime Security* (London; Routledge Curzon), 91–108.

Fell, Dafydd (2018) *Government and Politics in Taiwan*, 2nd edition (Oxon; Routledge).

Grove, Eric (2003) 'Conclusion: A Comprehensive Assessment of Taiwan's Sea Power', in Martin Edmonds and Michael M. Tsai (eds.), *Taiwan's Maritime Security* (London; Routledge Curzon), 169–174.

Hau, Pei-tsun (1995) *Hao Zongzhan Riji Zhong de Jingguo Xiansheng Wannian* (The Later Years of Mr JIANG Jingguo in the Diary of the Chief of the General Staff HAO Bocun), edited by Cora L. S. Wang (Taipei; Commonwealth Publishing).

Hau, Pei-tsun (2000) *Ba Nian Canmou Zongzhang Riji* (The Diary of the Eight Years as the Chief of the General Staff) (Taipei; Commonwealth Publishing).

Hau, Pei-tsun (2019) *HAU Pei-tsun Huiyilu* (HAU Pei-tsun's Memoir) (Taipei; Global Views-Commonwealth Publishing Group).

Hou, Kuang-hao (2020) 'Military Constructions of the Ocean on Post-War Taiwan in Three Phases', *Australian Journal of Maritime and Ocean Affairs*, 12(2), 61–82.

Hsiao, Hsu-tsen (2018) *Ba Nian Zhizheng Huiyilu* (A Memoir of the Eight Years of Governing) (Taipei City; Global Views-Commonwealth Publishing Group).

Huang, Guoliang (2019) Feidan Hui Zhuanwan? Liangan Kuangxiangchu Zhexie Ren ceng Paochu Jingtian Lei' (Turning Missiles? Fantasies across the Strait: Those People who Casted Thunder Bolts), *United Daily News*, 11 September 2019, accessible at https://theme.udn.com/theme/story/6773/4040779, accessed on 23 February 2020.

Huang, Huang-hsiung (2017) *Taiwan Guofang Biange: 1982–2016* (The Transformation of the Defence of Taiwan: 1982–2016) (Taipei; China Times Publishing).

Huang, Kwei-bo (2016) 'Taiwan's Foreign Policy and International Space', in Gunter Schubert (ed.), *Routledge Handbook of Contemporary Taiwan* (Oxon; Routledge), 465–481.

Hughes, Christopher R. (2016) 'National Identity', in Gunter Schubert (ed.), *Routledge Handbook of Contemporary Taiwan* (Oxon; Routledge), 153–168.

Huntington, Samuel (1993) *The Third Wave: Democratization in the Late Twentieth Century* (Norman; University of Oklahoma Press).

Jacobs, J. Bruce (2016) 'Taiwan during and after the Democratic Transition (1988–2016)', in Gunter Schubert (ed.), *Routledge Handbook of Contemporary Taiwan* (Oxon; Routledge), 36–67.

Lai, Ming-tang (1994) *LAI Ming-tang Xiansheng Fangtan Lu* (The Reminiscences of Mr. Ming-tang Lai), interviewed by Ming Lai (Taipei County; Academia Historica).

Lee, Pei-shan and Yun-han Chu (2016) 'Cross-Strait Economic Integration (1992-2015)', in Gunter Schubert (ed.), *Routledge Handbook of Contemporary Taiwan* (Oxon; Routledge), 410–425.

Lee, Teng-hui (1999) *Taiwan de Zhuzhang* (Taiwan's Assertions) (Taipei City; Yuan-Liu).

Lee, Teng-hui (2015) *Xin, Taiwan de Zhuzhang* (New, Taiwan's Assertions) (New Taipei City; Book Republic).

Lee, Zong-rong and Thung-hong Lin (2017) '"Taiwan Zhizao" de Jueqi yu Shiluo: Taiwan de Jingji Fazhan yu Jingji Shehuixue' (The Rise and Fall of Made in Taiwan (MIT): Taiwan's Development and the Legacy of Economic Sociology), in Zong-rong Lee and Thung-hong Lin (eds.), *Weijing de Qiji: Zhuanxing xhong de Taiwan Jingji yu Shehui* (Unfinished Miracle: Taiwan's Economy and Society in Transition) (Taipei City; Institute of Sociology, Academia Sinica), 1–43.

Leng, Tse-kang (2016) 'Taiwan and Economic Globalization', in Gunter Schubert (ed.), *Routledge Handbook of Contemporary Taiwan* (Oxon; Routledge), 233–250.

Li, Yuxi (2013) *LI Yuxi Xiansheng Koushu Lishi* (The Oral History of Mr LI Yuxi), recorded by Li Chang (Beijing; Jiuzhou Press).

Lin, Hsiao-ting (2016) *Yiwai de Guodu: JIANG Jieshi, Meiguo, yu Jindai Taiwan de Xingsu* (Accidental State: Chiang Kai-shek, the United States, and the Making of Taiwan), translated by Chung-hsien Huang (Taipei; Walkers Cultural Enterprise).

Liu, Yongqi (1994) *Haijun Kangzhan Shi* (The History of the Republic of China Navy's Resistance against the Japanese Invasion) (Taipei; Navy Command ROC).

Mahan, Alfred Thayer (2007 [1980]) *The Influence of Sea Power upon History 1660–1783* (Sioux Falls; NuVision Publications).

Michael M. Tsai (eds.), *Taiwan's Maritime Security* (London; Routledge Curzon), 137–149.

Ministry of National Defense (2000) *2000 National Defense Report, the Republic of China* (Taipei; Ministry of National Defense, Republic of China).

Ministry of National Defense (2006) *2006 National Defense Report, the Republic of China* (Taipei; Ministry of National Defense, Republic of China).

Ministry of National Defense (2009) *2009 Quadrennial Defense Review, the Republic of China*, English version (Taipei; Ministry of National Defense, Republic of China).

Ministry of National Defense (2011) *2011 National Defense Report, the Republic of China* (Taipei; Ministry of National Defense, Republic of China).

Ministry of National Defense (2013) *2013 Quadrennial Defense Review, the Republic of China*, English version (Taipei; Ministry of National Defense, Republic of China).

Murray, William (2008) 'Revisiting Taiwan's Defense Strategy', *Naval War College Review*, 61(3), article 3, accessible at https://digital-commons.usnwc.edu/nwc-review/vol61/iss3/3, accessed on 24 February 2020.

Przeworski, Adam (1991) *Democracy and the Market: Political and Economic Reforms in Eastern Europe and Latin America* (Cambridge; Cambridge University Press).

Steinberg, Philip E. (2001) *The Social Construction of the Ocean* (Cambridge; Cambridge University Press).

Stocker, Jeremy (2003) 'Missile Defense at Sea: Options for Taiwan', in Martin Edmonds and Michael M. Tsai (eds.), Taiwan's Maritime Security (London; Routledge Curzon), 137–149.

Su, Chi (2009) *Taiwan's Relations with Mainland China: A Tail Wagging Two Dogs* (London; Routledge).

Taylor, Jay (2000) *The Generalissimo's Son: Chiang Ching-kuo and the Revolutions in China and Taiwan* (Cambridge, Massachusetts; Harvard University Press).

Tsai, Michael M. (2003) 'Introduction', in Martin Edmonds and Michael M. Tsai (eds.), *Defending Taiwan: The Future Vision of Taiwan's Defence Policy and Military Strategy* (Oxon; Routledge), 1–9.

Wade, Robert (2004) *Governing the Market: Economic Theory and the Role of Government in East Asian Industrialization*, 2nd edition (Princeton; Princeton University Press).

Wang, Jenn-hwan, Zong-rong Lee, and Tsung-yuan Chen (2017) 'Taiwan Jingji Fazhan zhong de Guojia Juece: Lishi Huigu yu Lilun Zhanwang' (The State's Role in Taiwan's Economic Development: Historical and Theoretical Assessment), in Zong-rong Lee and Thung-hong Lin (eds.), *Weijing de Qiji: Zhuanxing xhong de Taiwan Jingji yu Shehui* (Unfinished Miracle: Taiwan's Economy and Society in Transition) (Taipei City; Institute of Sociology, Academia Sinica), 49–88.

Wu, Charles Chong-han (2018) 'Zhong-Mei Jingzhu xia de Taiwan Bixian Celü' (Taiwan's Hedging Strategy under the Influence of US-China Competition and Cooperation', *EurAmerica*, 48(4), 513–547.

Wu, Yu-shan (2016) 'Cross-Strait Dialogue and Policies', in Gunter Schubert (ed.), *Routledge Handbook of Contemporary Taiwan* (Oxon; Routledge), 393–409.

Xu, Xuehai (2002) 'XU Xuehai Xiansheng Fangwen Jilu' (An Interview with Mr XU Xuehai), in Li Chang, Shoucheng Wu, and Chin-lan Tseng (eds.) *Haijun Renwu Fangwen Jilu, Di Er Ji*, Volume 2: Collected Reminiscences of People in ROC Navy (Taipei; Institute of Modern History, Academia Sinica), 17–154.

Yang, Chih-heng (2003a) 'The Evolution and Adaptation of Taiwan's Military Strategy', in Martin Edmonds and Michael M. Tsai (eds.), *Defending Taiwan: The Future Vision of Taiwan's Defense Policy and Military Strategy* (London; Routledge), 51–66.

Yang, Chih-heng (2003b) 'The Contents and Goals of the ROCN's modernization', in Martin Edmonds and Michael M. Tsai (eds.), *Taiwan's Maritime Security* (London; Routledge Curzon), 79–90.

You, Ji (1999) *The Armed Forces of China* (St Leonards; Allen & Unwin).

5 The Social Constructions of the Ocean by the Economic Sources of Power in Modern Taiwan[1]

Introduction

This chapter aims to introduce how the primary marine-related economic powerholders in the society of modern Taiwan constructed the ocean after the central government of the Republic of China (ROC) withdrew to Taiwan in late 1949. Such an introduction not only presents the important historical experiences of how these powerholders used the ocean but also demonstrates one of the main theses of Philip E. Steinberg's (2001) territorial political economy: that the social construction of the ocean (or other geographical phenomena) can influence the material organisation of society established to use the ocean, which can, in turn, affect the natural condition, or the first nature according to Steinberg (2001), of the ocean.

Chapter 4 utilised a comparative method to show that the natural condition of the sea (that is, the Taiwan Strait) and the socio-material organisation of the military forces in Taiwan (mainly the ROC's grand and military strategies and the military balance between the armed forces of Taiwan and the Chinese mainland) resulted in corresponding constructions of the ocean in various phases by the ROC's military forces.

This chapter adopts a different method. It comprises two case studies, which are presented because they represent the most prominent types of economic powerholders related to the ocean before 2016: maritime shipping industries and fishing industries. Although the offshore wind industry has also been gaining in importance in Taiwan since 2016, its actual practices and effects still require more observation and should be explored on another occasion.

The first case study attempts to reveal how the maritime shipping industries constructed the ocean through examining their development. This case is relatively simple. The second case, however, provides more functions than the first. It describes and classifies the constructions of the ocean by the branches of modern Taiwan's fishery, as the first case does for Taiwan's maritime shipping industries and, also provides a narrative analysis of the influence of the constructions of the ocean by the fishery on the material organisation of society and, eventually, on the natural condition of the ocean.

DOI: 10.4324/9781003291343-5

In the discussion of the shipping case study, I first suggest that Taiwan's maritime shipping industries did not construct the ocean according to the Micronesia Model, given that their chief objective was to maximise profits and minimise the costs of their operations, which was not imperative to possess any portion of the ocean or sea routes. Second, although, as discussed in Chapter 2, the ROC incorporated elements of the Freitian and Grotian Variations at different moments as the situations demanded, the maritime shipping industries in the society of Taiwan adhered to the rationale of terrestrial industrial capitalists.

In other words, modern Taiwan's maritime shipping industries assisted their terrestrial cargo owners in reducing the costs of transport and distribution and, in turn, in increasing the amount of cargo and generating more profit. They had to ensure the smooth crossing of the ocean to deliver goods. They also had to undertake actions to reduce the costs of such a task. It can be argued that the construction of the ocean by the maritime shipping industries was in the style of the Indian Ocean Model.

In the second case study, I divide my narrative analysis into three parts. In the first, I argue that offshore aquaculture constructed the ocean consistent with the Seldenian Variation, and inland aquaculture constructed the ocean according to the Indian Ocean Model, as their practices and intentions of utilising the ocean have shown. In the second part, I posit that near-shore and near-sea fishery both constructed the ocean according to the Indian Ocean Model. However, after the 1990s, Taiwan's near-sea fishery absorbed the essence of the Seldenian Variation, demanding more protection by the state.

In the third part, I assert that modern Taiwan's deep-sea fishery constructed the ocean in the manner of the Indian Ocean Model prior to the late 1970s and early 1980s when many states in the world established their Exclusive Economic Zones (EEZs). After this period, I suggest that the deep-sea fishery of Taiwan gradually followed the rules and norms of the international community, incorporating attributes of the Seldenian Variation into its construction of the ocean. Moreover, the international community's regulations imposed by regional fishery management organisations incorporated components of the weak form of the Grotian Variation into Taiwan's deep-sea fishery, especially after the mid-2000s.

It is worth noting that the historical discussion of modern Taiwan's fishing industries also sheds light on Steinberg's (2001) theory of territorial political economy. The story of the fishery' development demonstrates that the social construction of the ocean can affect society's material organisation for utilising the ocean and influence the geographical conditions of the ocean. Taiwan's deep-sea fishing industry, together with its counterparts in other fishing countries, overharvested the living resources of the ocean, resulting in the formation and establishment of many international fishery management organisations, laws and regimes supported by scholars, environmental groups and states after the early 1990s, to protect global fishing resources from overfishing and depletion. Similarly, in the history of the development of Taiwan's near-shore fishery, the constructions of the ocean in the fashion of the Indian Ocean Model, especially their practice of using the ocean, first corresponded and facilitated the ROC's policies and laws to promote Taiwan's fishing industries. The near-shore and near-sea fishery then began to decline due to overfishing after the mid-1980s.

With the introduction to this chapter, it is worth briefly presenting its outline later. The next section deals with the social construction of the ocean by the maritime shipping industries, and the third section addresses the constructions by the fishing industries in the society of modern Taiwan. The final section concludes the chapter.

The Social Construction of the Ocean in the Development of Maritime Shipping Industries

The construction of the ocean by the maritime shipping industries in the society of modern Taiwan can be found in the development of several large maritime transport companies, including Chinese Maritime Transport (CMT), Taiwan Navigation (TN), Yang Ming Marine Transport (YMMT), Evergreen Marine (EM) and Wan Hai Lines (WHL). The development of these corporations involved policies and practices that aimed at maximising their competitiveness and their profitability. Those policies and practices also demonstrate how the firms have socially constructed the sea. This section adopts a thematic approach to analysing their development. Instead of providing a historical narrative (as I do in Chapter 4) and an overall survey of useful publications (as I do in Chapter 3), the thematic analysis in this section identifies some critical experiences of these maritime transport organisations. Through evaluating their experiences, the general construction of the ocean by the maritime shipping industries in modern Taiwan can be sketched out.

Ships

Historians of maritime transport and trade, for example, Pao-tsum Tai (2000) and Kuo-tung Chen (2013), usually trace the origins of the maritime shipping industries on Taiwan to the premodern era during the late Ming and early Qing periods (1616–1722). However, Taiwan's economy was mainly agricultural at that time, and maritime transport only grew in tandem with the increase of agricultural production and its population. As a result, Taiwan relied on the Chinese mainland for consumer goods and exported agricultural surplus and special native produce in exchange, including sugar, silk and deerskin. After 1862, tea, sugar and camphor became the primary goods for export (Chen 2013), as mentioned in Chapter 3.

High-sea sailing junks were used as the principal medium of maritime transport, and steamships gradually appeared in Northern Taiwan only after the 1860s. Foreign merchants, largely British, dominated industrialised maritime shipping at the time. After the First Sino-Japanese War in 1894, Japan annexed and colonised Taiwan, controlling its maritime shipping market and accelerating the employment of steamships (Tai 2000). Historians of Taiwan's maritime transport are right to focus on ships and shipped goods (cargoes), which have been critical for the survival and prosperity of maritime shipping industries in both traditional and modern societies.

Japanese steamships had been utilised for transport between Taiwan, the Chinese mainland, Japan and Southeast Asia. Bombardments by the United States (US) during the Second World War (1939–1945) inflicted severe damage on Taiwan's maritime shipping industries. Steamships were sunk, and only 96 smaller motorsailers were saved. After the war, the provincial government of Taiwan of the ROC was only able to dredge up some steamships and repair them. The first two salvaged steamships and 69 still usable motorsailers were assigned to the newly established TN in July 1946 (Tai 2000: 289–290). TN was initially founded jointly by the central and provincial governments; the central government held 40 per cent of its shares, whereas the provincial government owned 60 per cent (Tai 2000: 302–303).

A similar situation also occurred with China Merchants Steam Navigation (CMSN, transformed into YMMT in December 1972). Due to CMSN's inability to repay loans to the US and to affect a timely withdrawal to Taiwan, many of its ships were confiscated by the US and the People's Republic of China (PRC). Only 54 ships were left in 1950, even though CMSN was owned and supported by the ROC (Wang 2016: 183). Notably, the average age of the ships owned by CMSN and other maritime transport corporations on Taiwan was 19.04 years, and the average sailing speed of a ship was only 9.75 nautical miles per hour (Tai 2000: 351).

Under such conditions, TN and CMSN were unable to compete worldwide. They had to muddle through operational difficulties with the state's help (Tai 2000) as well as expand and renew their fleets. Building a new and necessary ship was an expensive endeavour for them. For example, to construct a needed large bulk carrier of 28,800 metric tonnes of deadweight (deadweight tonnage, DWT), TN spent approximately 5.2 million US dollars and borrowed roughly 26 million US dollars in the early 1970s (Chi 2013: 206).

However, borrowing a large amount of money to acquire necessary ships could cause serious financial problems for a maritime transport company. In TN's case, even in the mid-1970s, it still had to pay back 1.1 billion New Taiwan Dollars (NTDs) and rely on short-term loans, which amounted to 240 million NTDs, to continue its operations. TN had to repay more than 30 million NTDs of annual interest. Even in 1976, for example, CHI Mengbin (the then director of the governing board of TN) met with JIANG Jingguo (the then premier of the ROC) in person to secure the required amount of money from the central government (Chi 2013: 208–209).

The reason for the ROC's reorganisation of TN and CMSN was their financial plight. Understandably, TN and CMSN began their operations in Taiwan in difficult times. Their lack of adequate ships caused financial, organisational and operational challenges for them, impeding their development. For example, as a former harbour master, LI Lianchi recalls that when he kindly advised the director of the board of the CMSN in 1969, CAO Zhongchou, on the issue of containerisation of ships, Cao regrettably remarked that they lacked funds for shipbuilding. CMSN could only construct container carriers ten years later after it had been transformed into YMMT (Li 1998).

Another example can be found in the initial development of EM. ZHANG Rongfa (also known as CHANG Yung-fa) had to borrow money from anywhere he could because of the poor quality of his ship, Central Trust. As he states in this memoir, he considers himself lucky that he received financial assistance from a Japanese commercial corporation, Marubeni. Marubeni's continuous financial support allowed EM to build new ships and grow (Chang 1997).

Possessing sufficient ships of good quality is indeed important for maritime transport corporations. Possessing the right types of ships is critical as well. According to LI Lianchi, he was the first person to introduce containers and container carriers to the chair of the board of directors of CMT, DONG Haoyun. As early as March 1969, when Li visited Port Jackson Bay in Sydney, Australia, he was determined to develop Kaohsiung harbour into a world-class container port. He informed Dong that over 65 per cent of the maritime transport between the US and Europe was containerised and that Japan's Showa Line had already deployed four container carriers on the Pacific Ocean (Li 1998: 79–80).

DONG Haoyun established CMT in Shanghai in March 1940. Fortunate for him, the Second World War did not destroy his financial foundation. After the war, he was able to cooperate with the Japanese shipbuilding industry to construct new ships and develop his maritime shipping enterprises. In September 1969, Dong reorganised one of his Hong Kong-based companies into the Orient Overseas Container Line and transformed seven Victory-class bulk carriers into container carriers. Its first container carrier set sail from Kaohsiung as early as November 1969 (Wang 2016: 121–124).

Having secured sufficient financial supply from Marubeni, ZHANG Rongfa started market surveys and research of containerisation in October 1972. He chose the shipping line between the Far East and the Eastern US and built four container carriers of 646, Twenty-foot Equivalent Units (TEUs). As a result, the first EM full container carrier, Ever Spring, set sail on the Far East-Eastern US sea route in July 1975 (Chang 1997: 137–138, 146). Zhang understood that the containerisation of maritime transport was a global trend.

Containers can adequately protect the cargo inside, whereas shipment on a bulker usually suffers a loss of 5 to 10 per cent of the total freight. Furthermore, using containers can also significantly reduce the time of shipping (Chang 1997: 129–130). A container ship can manage to load and discharge cargo in 10 hours that requires three days to complete by a bulk ship.

On average, when employing a container carrier, a maritime transport company can save 80 per cent of the time in port and 65 per cent of harbour fees consumed by a bulker (Wang 2016: 72).

In his memoir, ZHANG Rongfa states that the development of commercial and industrial activities around the world since the end of the Second World War generated an immense need for maritime transport of raw materials and industrial products. Such an immense need resulted in an increase of ship tonnage and the capacity of ports. Port traffic conditions around the world were deteriorating. Thus, containerisation provided a solution.

ZHANG Rongfa believed that he had to catch up with such a trend in the US and Japan for EM to further develop. 'Other maritime transport companies had already provided containerised shipping for approximately five years, If I had not taken any action at the time, it would have been very difficult for me to secure sufficient freight, needless to say competing against others' (Chang 1997: 133).

In contrast, YMMT had to wait until June 1978 to launch its first containerised shipping service. Nonetheless, it also strove to catch up with the trend of containerisation once its financial problem had been by and large handled and controlled. Within ten years, between 1978 and 1988, CAO Zhongchou and CHI Mengbin (the then Chief Executive Officer of YMMT) supported and ordered the construction of 11 container ships (Chi 2013: 224–225).

Additionally, and back to the case of EM, ZHANG Rongfa had begun to pursue containerisation when the First Oil Crisis (1973–1974) struck the global economy. With the crisis 'came the quadrupling of oil prices' (Wade 2004: 96). 'Huge amounts of hard currency accumulated in the treasuries of the Middle East oil-exporting countries, which in turn invested them around the world … High inflation plagued the United States and Europe for years afterward' (Goldstein and Pevehouse 2013: 405). On Taiwan, inflation accelerated in 1974, and the 'prices of oil and other energy-related products were raised even more—gasoline by 85 percent and electricity by nearly 80 percent' (Wade 2004: 97). Nevertheless, Zhang had already contrived a plan to cope with a possible economic downturn.

When ZHANG Rongfa was preparing to deploy container carriers entirely on the Far East-Eastern US shipping line in 1973, he primarily took market demand and corporate operation efficiency (including the number of crew and fuel efficiency) into consideration. He acknowledged that containerised shipping during that time had still not become the mainstream approach of maritime transport. Therefore, he only built four carriers of 646 TEUs for the line (Chang 1997: 137–138, 148).

Not only did these four carriers consume cheaper diesel fuel, but also each was designed to complete fast hull enlargement for carrying more containers, should market demand increase. Furthermore, when he aimed to containerise EM's Far East-Middle and South America liner service, he ordered the installation of four container cranes on each container ship sailing on this route to unload containers in ports where gantry cranes were not available

(Chang 1997:153). He proudly states that his plan allowed EM to manage the oil crises, and he always offered a special design for every new type of ship that EM ordered to be built (Chang 1997, 147–150).

In the earlier passages, I have used the historical experiences of some important maritime transport companies to demonstrate how critical ships were for them. Without sufficient, suitable ships of good quality (much more than mere seaworthiness), a maritime transport company could not survive international competitions. For TN and CMSC, they had to borrow money and became dependent on the state's fiscal supply. Even for a private company, EM relied heavily in the beginning on loans from Marubeni to purchase suitable ships for its consolidation and development. For the more fortunate CMT, its adequate fleet was able to develop, compete against other international merchant navies after Second World War and became the first Chinese containerised maritime transport enterprise. Furthermore, deploying container carriers with intelligent design and mindful planning in due course lead maritime shipping industries to prosperity, as shown in EM's case, and also in YMMT's and TN's attempts at revitalisation. Below I shall discuss how critical cargoes are for maritime shipping industries.

Cargoes

That shipping adequate cargoes is crucial for any merchant navy is a straightforward proposition. It should not be surprising to see that ZHANG Rongfa took creative actions to secure their sources of cargo. His creativity began with acquiring suitable types of ships for their cargoes, which permitted securing and attracting more cargoes. Using container carriers after the mid-1970s, as mentioned earlier, is an example. Earlier, when Zhang was the Chief Executive Officer of Xintai maritime transport company and shipping goods between Taiwan and Japan, he ordered the construction of a six thousand DWT general cargo ship for carrying wood (Chang 1997: 37–38).

This new ship was specially designed to transport wood products and using this ship became a novel practice in Taiwan's maritime transport industry. At the time, the industry often employed US Liberty-type wartime supply ships to carry wood. The size of each of the ships was usually ten thousand DWT. Because of the cumbersome volume of wood and the unsuited design of cargo holds, this type of ship was not an efficient choice for shipping wood between Japan and Taiwan. Zhang's choice could lower the cost of transportation and thus attract more customers for the company (Chang 1997: 37–38).

Another example of building ships suited for attracting and carrying more special cargoes can be found in the experience of TN. Bananas had been an important agricultural product exported to Japan through Taiwan-Japan maritime shipping lines since October 1950 (Tai 2000: 273). TN was a major company responsible for shipping bananas on this line. Despite its financial difficulty, TN still decided to purchase four refrigerated cargo ships in the early 1960s to decelerate the ripening process of bananas on board. Consequently, the amount of freight for TN increased, which reached its peak between 1968 and 1970. TN carried out 83 and 85 voyages in 1968 and 1969, respectively, and in 1970 conducted even more than 100 (Tai 2000: 298).

The second bold and inspired move of Zhang Rongfa was to undertake deep-sea liner shipping in August 1969, when the global maritime shipping environment did not favour such a move by a small-scale private company in Taiwan. This was because almost all deep-sea liner shipping services at the time had been monopolised by certain freight conferences. Zhang's first deep-sea liner service was between the Far East and the Middle East (Chang 1997: 73–77).

This route had been dominated by the freight conference comprising NYK Line, Mitsui OSK Lines, Maersk and some other large maritime transport corporations of a similar scale. Zhang believed that near-sea maritime shipping had become crowded and was getting more competitive. For a new, small EM to grow, it had to embark on deep-sea liner shipping to transport more cargoes and earn more profits (Chang 1997: 73–77).

As mentioned above, freight conferences had occupied the deep-sea shipping market. Therefore, Zhang Rongfa's third inventive action was to win over support from owners of cargoes, especially on Taiwan. Zhang realised that the owners on Taiwan could successfully secure orders for their goods in the Middle East market, but it was difficult for them to acquire sufficient space in cargo holds. Furthermore, because of the monopoly of maritime shipping services by the freight conference, the quality of the shipping was not satisfactory, and freightage was very expensive for the owners (Chang 1997: 77–78).

Thus, under Zhang's lead, EM approached these owners spontaneously, offered adequate cargo space for their goods with attractive, lower freight costs (about 15 per cent lower than that of the freight conference; Tai 2000: 428) and a better quality of service. Furthermore, Zhang and EM helped the owners of goods on Taiwan and in the Middle East increase the market share of their goods (Chang 1997: 77–78; Wang 2016: 148–149). This strengthened the relationships between EM and the owners and allowed EM to gain more customers (Wang 2016: 157–158).

A closely related issue, which demonstrates the fourth innovative policy, that EM attempted to deal with was establishing more deep-sea liner shipping routes to form an East-West two-way around-the-world shipping line in 1984 (Chang 1997: 222). The history of developing those deep-sea liner shipping services between various areas and how Zhang and EM overcame the hindrance of freight conferences do not need to be presented here. Zhang introduces it in his autobiography, but he also mentions that it was necessary to form this around-the-world line to reduce the repetition of port entering, save more time and money of transport and make the delivery and arrangement of containers more flexible (Chang 1997: 205).

Zhang Rongfa explains in his autobiography the factors he had to seriously consider when planning to operate the around-the-world line. The issues he had to contemplate included the operations procedures at each base along the line, how those procedures could be smoothly connected, the coordination with ports, the management of containers, the deployment of fleets, the arrangement and acquisition of cargoes and the connection and flow of the information related to operations around the world (Chang 1997: 209; Tai 2000: 429–431).

The most important decision that Zhang Rongfa made was to unify the type of ships used for two-way around-the-world liner shipping. At the beginning, Zhang deployed 16 G-type container carriers with a capacity of 2,728 TEUs in 1984. Once the operation of the around-the-world line was consolidated, 24 more G-type carriers were added (Chang 1997: 212).

Furthermore, he kept renewing the vessels on the line with more advanced types of ships to maintain the competitiveness of EM's service. Notably, Zhang always maintained a unified type of ship operating along the lines. This allowed EM to implement a stable operating system and made the management of containers and space for transport more convenient and coherent. Such an approach to operating the around-the-world line saved more than 58 million US dollars annually in the 1980s (Tai 2000: 430).

On the contrary, although YMMT creatively adopted and benefited from its pendulum sailing route for its containerised maritime shipping between North America, Japan, Taiwan and Europe in 1989, this method lasted for only three years. It ceased because YMMT did not deploy the same type of ships. Deploying different types of ships with varying cargo space resulted in inaccurate estimation of cargoes, spaces and freight forwarding (Wang 2016: 205–206).

Although the pendulum shipping service stopped in 1992, YMMT began to cooperate with Korea's Hanjin Shipping for joint operations in 1990 so that these two companies could fully utilise their cargo space. According to their agreement, Hanjin deployed five ships and YMMT four to transport cargoes to the eastern US Previously, YMMT had to sail nine ships to the same destination. Now, the extra five ships could be deployed on the Far East-Western US line, allowing YMMT to begin earning profits (Wang 2016: 209–210).

The success of the joint operation between YMMT and Hanjin led to an alliance between YMMT and WHL in 1992. The alliance with WHL, which specialised in near-sea containerised maritime transport, permitted YMMT to initiate its near-sea shipping. In 1994, YMMT cooperated with NYK Line and Le Groupe CMA CGM to operate the Far East-Mediterranean and South Europe Line. In 1995, YMMT cooperated with China Ocean Shipping Company (COSCO). YMMT even formed a CKY alliance with COSCO and Kawasaki Kisen Kaisha in 1996. The CKY alliance was later joined by Hanjin in 1997 and became the CKYH alliance. Through joint operations of the CKYH alliance, YMMT was able to expand its shipping lines around the world. In 2014, EM also joined CKYH, becoming the CKYHE alliance. YMMT's joint operation and alliance strategy was successful, enabling YMMT to appropriately and efficiently allocate its shipping capacity and acquire needed cargo (Wang 2016: 211–213).

Thus far, I have discussed several creative approaches devised by maritime shipping corporations to enlarge their sources of cargo. Reducing time and the cost of transport by containerisation was one way. Deploying ships suited to carry specific cargoes was also effective. Exploring and exploiting the market that demanded more deep-sea liner shipping services, for example, the Far East-Middle East line in the late 1960s and early 1970s, also helped acquire goods to be transported. Furthermore, EM established an around-the-world shipping line. Although YMMT's attempt at pendulum services in the late 1980s and early 1990s failed, it gradually became the CHYHE maritime shipping alliance. EM's around-the-world container liner shipping and YMMT's establishment of maritime alliances allowed both of them to achieve operational efficiencies, reduce costs, increase cargoes and generate more profits.

Logistics Support on Land

Logistics support for maritime transport is complicated. This subsection focuses on only two of the most important issues of the logistics system of any maritime transport corporation, both of which are pointed out by Zhang Rongfa of EM and CHI Mengbin of YMMT in their memoirs. A discussion of these two issues, together with the above thematic overview of the development of some important maritime transport companies, demonstrates how the maritime shipping industries in the society of modern Taiwan constructed the ocean.

The first issue mentioned by both ZHANG Rongfa and CHI Mengbin is personnel recruitment and human resource management. This issue was an internal managerial matter of each maritime transport corporation but was crucial for the survival and development of the organisation. CHI Mengbin of YMMT also established the requirement that recruitment take place by state examination through the ROC's Examination Yuan and the Ministry of Transport in 1978, the first year when he became the company's Chief Executive Officer. He believed that this would recruit more young people with genuine maritime expertise into YMMT.

Zhang was particularly concerned with this issue. When he was in Xintai maritime transport company in the early 1960s, he was aware of the problem of hiring persons who were introduced and recommended by central or local government officials or even by people in national security units. The performance of such staff could hardly be considered satisfactory, and many of them simply idled around (Chang 1997: 46). Therefore, when Zhang founded EM on his own, he established a policy of personnel recruitment based only on merit. The evaluation of candidates comprised two sessions: first, written examinations, and second, interviews by experienced managers. EM normally held evaluations for young graduates because Zhang believed they were the most amenable to EM's own organisational culture and cultivation as they had just graduated from schools and colleges (Chang 1997).

As Zhang's mentions in his autobiography, the core of EM's internal education programme adopted a master-to-pupil approach. A senior staff of a unit in EM would be assigned to teach, train and discipline an individual or several new recruits of the unit according to their missions and EM's organisational culture and values (Chang 1997). Notably, Zhang is that he created a ship-shore rotation training scheme for seafarers of EM, the purpose of which was to provide the opportunity to work onshore after a certain period of time on board. In addition, the scheme allowed seafarers more time to stay with their families and a chance to cooperate with their land-based colleagues. Under this scheme, both seafarers and land-based colleagues could learn different perspectives on their roles in EM and better understand the processes and procedures of their tasks (Chang 1997).

The second issue that both CHI Mengbin and ZHANG Rongfa mention in their memoirs is the building of a logistical network for containerised maritime shipping. Chi mentions that he worked hard with his colleagues to establish a network of liner shipping agents in various important port cities in Europe when YMMT was striving to increase its market share of the Far East-Europe liner shipping market. For example, he led a team to visit London in October 1980 and discussed cooperation between YMMT and Johnson & Stevens Agencies. Johnson & Stevens became YMMT's chief shipping agent in Europe with favourable terms for the Taiwan's firm. Furthermore, YMMT also appointed Intersea SRL in Italy, Cie Charles Le Borgne in France, Catalana Del Mar SA in Spain, FA Voigt & Co BV in the Netherlands, Unimar Gesellschaft Mbh in Germany and Muller-Thomsen Antwerpen NV in Belgium as shipping agents (Chi 2013: 229–231). In this way, a comprehensive network of containerised liner shipping agents was formed.

Here I will not go into the EM network of shipping agents and YMMT's network in other regions. Instead, I will introduce three of EM's important subsidiary companies in Taiwan. Zhang Rongfa explains his rationale behind establishing Evergreen Heavy Industrial Corporation in 1982 and Evergreen Container Terminal Corporation in 1983. As mentioned earlier, EM initiated its containerised deep-sea liner shipping service in July 1975. As its fleet of container carriers grew larger as the coverage of its service expanded, it became necessary for EM to construct its own facilities to build, maintain and repair its containers.

EM's container construction and maintenance factory was first established in Kaohsiung. Because geographical limitations restricted the available space, EM later developed a port in Keelung and also had to build a container terminal inland. Thus, the largest inland container terminal in the Far East was installed in Taoyuan in 1983. Together with an expanded fleet of inland container haulage operated by Everglory Transport Corporation (which was established in September 1973), these subsidiary companies allowed EM to control the quality and efficiency of almost every aspect of its maritime shipping operations (Chang 1997: 210–212).

In fact, CMT had constructed the first inland container terminal on Taiwan to solve the problem of space limitation in Taiwan's container ports in 1971.

The inland container terminal made it possible for customs officials to complete their customs clearance procedures, and cargo owners could load and unload containers on site in the terminal. CMT also established its own inland container haulage fleet in 1972 to provide the first door-to-door shipping service. Additionally, CMT followed the state's policy to foster the indigenous shipping industry on Taiwan and provided financial support to invest in the state-owned China Ship Building Corporation in 1973. When Taiwan's maritime shipping industries were growing in the 1970s (Lee 1992: 245–247), CMT's decision to expand its related businesses on land consolidated its importance in Taiwan's economy and secured its pathway to further development after it transferred its containerised maritime shipping service to the Orient Overseas Container Line in Hong Kong in the mid-1980s (Wang 2016: 127–128, 131–132).

Uniglory Marine Corporation, established in August 1984, is the third EM subsidiary discussed here. Initially, Uniglory Marine was responsible for near-sea maritime shipping, both liner and non-liner transport services. This mission was assigned because EM was mainly in charge of deep-sea containerised liner shipping. EM would only transport cargoes between major ports in various continents, whereas Uniglory transferred goods shipped by EM from mainly the East-West deep-sea lines to intra-regional harbours. Uniglory generally sailed in a North-to-South pattern, helping EM form a maritime shipping network in East Asia. This pattern became particularly prominent in the 1990s (Chi 2013: 220; Tai 2000: 436–440).

The earlier discussion illustrates the important policies that the chief executives of large maritime shipping enterprises would contemplate and implement, in addition to handling the normal maritime affairs related to operations and ships. First, talent and expertise were particularly valued by those chief executives, for example ZHANG Rongfa and CHI Mengbin. Second, controlling the whole process of containerised maritime shipping to enhance quality and reduce transportation costs was the driving reason for the large maritime shipping companies extending their business inland. Container terminals and terrestrial container haulage fleets became useful, as shown in the cases of EM and CMT. Third, deep-sea container liner shipping services could be supplemented by near-sea maritime transport to transfer cargoes to minor intra-regional harbours. Such a complementary shipping service was inevitable as container carriers grew larger, connecting fewer major ports and as container shipping became more popular. EM's subsidiary Uniglory carried out this function. The alliance between WHL and YMMT mentioned earlier also provided this service.

The Social Construction of the Ocean by Maritime Shipping Industries

YMMT, EM and WHL continued their process of growth and transformation throughout and after the 1990s. YMMT completed its privatisation in 1996. EM established EVA Air in 1989 and underwent a thorough reorganisation in the 1990s and 2000s. EM and EVA Air developed into the giant Evergreen Group. WHL became the fourth largest maritime transport company on Taiwan after 1984 (Wang 2016: 250). In Alphaliner's world top 100 operated fleets in June 2011, YMMT ranked 15th, EM 4th and WHL 21st (Alphaliner 2011). TN and CMT also dovetailed in their market; they transformed their organisations and embarked on customised shipping services after the 1990s. TN also completed its privatisation in 1998.

I do not, however, introduce the history of their development and transformation. The purpose of the discussion in this section, thus far, is sufficient to demonstrate the construction of the ocean formulated by modern Taiwan's maritime shipping industries. I have categorised the uses of the ocean and business practices that were considered important by the presidents of some major maritime transport companies by several themes. I have analysed their employment of customised ships, collection of profitable cargoes and establishment of terrestrial logistics. All in all, the maritime shipping industries aimed to maximise profits, minimise costs and maintain the competitiveness of their operations and companies.

This assessment, in effect, has already excluded the categorisation of the Micronesia Model. Modern Taiwan's maritime shipping industries did not view the ocean as a resource reservoir. Their profits were earned by providing the service of transporting goods for customers on land, even though their approach to transport was crossing the surface of the ocean. Moreover, the shipping industries did not consider the seas as possessable. It would be in their interest not to endorse any claim to ownership of a certain portion of the ocean, which might result in blocking certain sailing routes on which the shipping industries on modern Taiwan, as well as maritime industries in other places in the world, depended.

Smooth movement not only referred to effective cargo transportation but also efficient delivery. For both maritime shipping providers on modern Taiwan and cargo owners worldwide, delivery efficiency meant fast, timely, cheap and quality shipping services. A lower rate of damaged cargoes during transportation meant lower costs. Containerisation, which protected the cargoes they held, provided a solution to this problem. With appropriate facilities, container carriers could load and unload faster than traditional general cargo ships. When international trade grew, the size of container carriers also increased to cope with the need to circulate ever larger volumes of goods (Steinberg 2001).

Some inland container haulage fleets and terminals operated by Taiwan's maritime transport corporations even evolved into larger terrestrial storage, logistics and transportation service corporations, for example, Evergreen International Storage and Transport Corporation. Initially, inland container haulage fleets, however, were deployed to ensure that each step of the channelled circulation of goods could be appropriately and efficiently controlled. Moreover, to overcome terrestrial spatial constraints on container ports, especially Keelung Port, large maritime shipping corporations on Taiwan built their own container terminals, which became convenient both for cargo owners and for customs to operate.

None of these wealth-generating approaches adopted by maritime transport companies was directly related to the ocean *per se*. Whether or not the ocean could provide living and non-living resources or could be owned and territorialised were not the main concern of the maritime shipping industries. The maritime transport of goods only utilised the surface of the ocean (so long as the right of navigation was not infringed). It can thus be argued that maritime shipping industries on modern Taiwan did not construct the ocean in accordance with the Micronesia Model.

The term maritime circulation derived from Steinberg's (2001: 73) phrase 'closed, channeled paths of circulation', can now be explicated. It simply refers to controlling overseas trade and the transportation of goods through certain chosen routes. When Steinberg (2001) uses this term in his monograph, however, it is employed in the context of the European mercantilist era, roughly between 1450 and 1760. At that time, channelled circulation indicated the intention of mercantilist capitalists and sovereigns to accumulate precious metals through,

> ensuring that the entire product of a region found its way into their ships, that these ships found their way to their designated destination, and that distribution within the land of destination also remained under their control. National policy supported these closed networks as means toward minimizing the export of precious metals. Indeed, the genius of the strongest mercantilist powers lay in their ability to triangulate this trade in such a way as to generate a net importation of precious metals.
>
> (Steinberg 2001: 73–74)

Consequently, 'in a system in which economic power was based upon controlling discrete channels of trade, the surface upon which trade was carried out (the ocean) would emerge as a site for exercising power', as Steinberg (2001: 74) points out.

For Steinberg (2001), the construction of the ocean in the European mercantilist era by the general marine user groups was in line with the Freitian Variation of the Mediterranean Sea Model. As discussed in Chapter 2, the ROC between the early 1950s and late 1980s also absorbed elements of marine construction that were consistent with the Freitian Variation. I have demonstrated the state's construction through an examination of its stipulated maritime laws in that chapter.

The ROC also implemented policies to build domestic vessels with domestic shipbuilding yards and to transport domestic products with domestic vessels. Before TN was privatised and CMSN was transformed into YMMT, these two companies were requested to transport civil and military necessities. For example, the state demanded that TN provide transportation for the military six times in 1946. Throughout the entire year 1959, TN had to assign one ship solely for military transport (Tai 2000: 301).

In the case of CMSN, 80 per cent of the withdrawal mission from Shanghai that took place in 1949 during the Chinese Nationalist-Communist Civil War (1946–1949) was accomplished by this state-owned corporation. It then became responsible for the transport of the retreats from Hainan, Zhoushan and Dachen Islands. When the ROC consolidated its rule on Taiwan, CSMN still assisted the state and provided services for military needs without any remuneration. For example, in May 1953, it offered transport services to the armed forces that involved 10 ships for 12 days, incurring losses of approximately 100 thousand NTDs (Tai 2000: 352).

The ROC's exploitation of CMSN was not limited to services for the military. CMSN was also required to make compulsory shipments of indigenous agricultural products for export, especially after the outbreak of the Korean War (1950–1953). The demand for shipping related to the war increased, yet the maritime transport of Taiwan's agricultural produce made freight much less attractive to foreign maritime shipping companies at the time (Lee 1992: 238).

Additionally, for national security reasons, CMSN was requested by the state to transport bulk cargoes, grain, minerals and fossil fuels. Taiwan had (and still has) to import these types of commodities (Chi 2013: 220). The need to import required and strategic bulk goods was a reason for the ROC's rejection of YMMT's request to acquire seven container carriers and forcing it, instead, to obtain three Panamax bulk carriers of 66 thousand DWT in the late 1970s and early 1980s (Chi 2013: 234–235).

In addition to the evolution of the ROC's maritime laws discussed in Chapter 2, the policies of maritime transport of civilian and military necessities mentioned above allow readers to realise that the ROC, between the 1950s and 1980s, also constructed the ocean in the style of the Freitian Variation regarding maritime shipping. To support the development and foster a sound industrial capitalist economy on modern Taiwan, the state made a great effort to regulate and promote maritime trade and transport. Requesting state-owned maritime shipping companies to transport military equipment and personnel without due compensation was detrimental to the companies. This type of request, along with securing a guaranteed means to import and export necessary goods across the ocean, facilitated the ROC's provision of social stability for Taiwan's economy to grow. The state's construction of the ocean regarding maritime shipping during this period could properly be categorised into the Freitian Variation.

However, the construction of the ocean by modern Taiwan's maritime shipping industries could hardly be classified in the same category. CHI Mengbin of YMMT complained that the state's decision to compel his company to obtain bulk carriers made YMMT lose money on the ships and delay its containerisation (Chi 2013). A major factor that caused TN and CSMN to be in a deficit and to carry debt was the state's requirement for compulsory transport services for the military (Tai 2000). That is, the state's intervention in the practices of maritime shipping was not necessarily in the interests of the related companies.

Furthermore, the policies and actions of maritime shipping industries (including inland container haulage and terminal services) undertaken by the aforementioned large corporations on modern Taiwan did not require the ROC's stewardship of the trade sea routes. Although the state attempted to implement policies for closing and channelling the circulation of goods to facilitate the development of Taiwan's industrial capitalist economy, Taiwan's large maritime transport corporations did not show the same inclination.

These corporations did strive to increase the amount of their cargo. However, their rationale underlying their operations was to maximise profits and minimise costs. To monopolise the channelled circulation was not their goal, especially if the monopoly would result in inefficiencies and a diminishing marginal utility. This was one of the reasons that EM, YMMT and WHL all employed different strategies for their maritime shipping services. For example, as shown in EM's and YMMT's endeavours to offer lower freight charges when they were challenging the monopolies of freight conferences, they understood and followed the rationale of terrestrial industrial capitalists. That is, lower costs of production and distribution meant greater competitiveness, with other conditions being equal. EM's and YMMT's efforts to offer lower freight charges than that of other freight conferences could help their terrestrial capitalist customers increase the competitiveness of their products and hence increase EM's and YMMT's cargoes and profits.

The notion that reducing the costs of distribution could raise the competitiveness of goods is consistent with Steinberg's (2001) analysis of the spatial logic of industrial capitalism: 'The industrial era brought about the classic "capitalist spatiality ... one in which dual tendencies towards capital mobility and capital fixity were associated with capitalist alternatively investing (or "filling") and disinvesting in (or "emptying") discrete location on the earth's surface' (Steinberg 2001: 110). The key to deciding where and when to fill and empty capital to and from a developable inland location is the cost of production and distribution imposed by producing and shipping products on and from the location.

The ocean, according to this spatial logic, represents a dispatch distance that must be eliminated and a surface across which products are to be transported and delivered. This is a construction of the ocean according to the Indian Ocean Model. Modern Taiwan's maritime shipping industries, which strove to attract more cargoes and earn profits by adhering to the rationale and spatial logic of industrial capitalism, also constructed the ocean in accordance with the Indian Ocean Model, rather than the Freitian Variation.

To summarise this section, I have discussed several important themes considered important for the survival, consolidation and development of modern Taiwan's maritime transport corporations. Customised ships of good quality, such as a new container carrier for deep-sea liner shipping or a modern reefer ship for fresh fruit delivery, were crucial to raise and maintain the competitiveness of the corporations. Obtaining enough cargoes to ship was certainly critical, too. Therefore, maritime transport corporations, especially the large ones, often evolved into sets of connected maritime shipping enterprises, controlling almost every step of their door-to-door transport service of goods for industrial capitalists on land. For example, YMMT operated its inland container terminal, and EM grew into the Evergreen Group, not only offering deep-sea container liner shipping but also formulating a near-sea maritime transfer network. All these endeavours sought to maximise their profits and minimise their costs, thus following and enabling the logic of spatiality of industrial capitalism. Thus, it can be argued that the maritime shipping industries on modern Taiwan constructed the ocean in accordance with the Indian Ocean Model.

The Social Constructions of the Ocean in the Development of Fishing Industries

The fishing industries on modern Taiwan constructed the ocean in various ways. As in the case of the ideological powerholders, different intentions of using the ocean led to different representations and, in turn, constructions of the ocean. Each branch of the fishing industry constructed the ocean variously in accordance with how they utilised the ocean. I have demonstrated that how people do, intend to, or urge others to use the ocean affects how they construct the ocean in Chapter 2 and 3.

As discussed in Chapter 2, the ROC absorbed elements of various types of constructions of the ocean from the international community. Between the 1970s and 1990s, regarding the fisheries, the state constructed the ocean according to the Seldenian Variation. However, the constructions of the ocean by the fishing industries in the society of modern Taiwan have deviated from that of the state, except in cases of requiring its protection and stewardship. In this section, I follow the usual classification of Taiwan's fishing industries: deep-sea fishery, near-sea and near-shore fishery and offshore and inland aquaculture (Chen, Bao-ji 2012: 4–5; Hu 2007: 7–8). I shall make the following points.

First, inland aquaculture constructed the ocean in the style of the Indian Ocean Model, whereas the offshore aquaculture industry constructed the ocean in the fashion of the Seldenian Variation. Second, the near-shore fishing industry constructed the ocean consistent with the Indian Ocean Model and remained so up until 2016. The near-sea fishing industry initially constructed the ocean in accordance with the Indian Ocean Model but later, especially after the 1990s, absorbed elements of the Seldenian Variation, as they began to demand more state protection over and stewardship of their fishing grounds that had been situated in overlapping EEZs. Third, the deep-sea fishing industry constructed the ocean corresponding to the Indian Ocean Model. However, under pressure from the international community, it also absorbed elements of the Seldenian Variation after the late 1970s and of the Grotian Variation after the mid-2000s.

It is worth noting that the fishing industries on modern Taiwan constructed the ocean differently from the state's construction of the ocean, even though these industries occasionally depended on the state's financial support and legal regulation. Their uses of the ocean were not coherent with the state's intention to use and regulate the uses of the ocean. Below, I first introduce the social construction of the ocean by modern Taiwan's aquaculture through the trajectory of its development.

Inland and Offshore Aquacultures

According to the earliest records, aquaculture emerged in Taiwan during the late Ming dynasty, roughly in 1600. Initially, aquaculture technology spread from Fujian and Guangdong Provinces of the Chinese mainland (Shi 2012: 121). During Japanese colonial domination, aquaculture played a minor, supplementary role in the whole Japanese imperial economic system. Taiwan's agriculture, including aquaculture and other fishing industries, was designed and required to provide a surplus for industrialised Japan. The annual amount of aquaculture production in Taiwan at that time was approximately 20 thousand tonnes (Shi 2012: 122).

According to SHI Shenglong, a former director of the Aquaculture Fisheries Division of the Fisheries Agency of the Council of Agriculture, aquaculture after 1949 underwent several phases. First, between 1951 and 1970, the technical foundation of the aquaculture industry was established with the state's scientific and financial help. Although the amount of production only increased to more than 70 thousand tonnes per year, the technologies of cultivating white amurs, silver carps, eels, tilapias, giant tiger prawns and scalloped spiny lobsters were developed. This technological foundation allowed Taiwan's aquaculture to grow rapidly in the second phase, between 1971 and 1980 (Shi 2012: 122).

During the second phase in the 1970s, the size of Taiwan's aquaculture grew approximately 17 thousand hectares and its production to 95 thousand tonnes. Notably, the production of freshwater aquaculture increased approximately 7.6 thousand tonnes, whereas the inland salt-water aquaculture grew to only 2.1 thousand tonnes (Shi 2012: 122). Offshore aquaculture still lagged, even though it was also growing in the second phase, representing approximately 11.4 per cent of Taiwan's total aquaculture industry (Su 2007: 88).

The third phase was in the 1980s and 1987 saw the apex of Taiwan's aquaculture production at more than 300 thousand tonnes. However, giant tiger prawns, eels and tilapias were the main products of the aquaculture industry between 1980 and 1990. Although many villages of aquaculture were in proximity to the sea, the industry pumped fresh underground water for its operations. Such an approach led to serious land subsidence, especially in the coastal areas southwest of Taiwan Island (Shi 2012: 123–124).

In the fourth phase, subsidence resulted from the state's efforts to encourage and expand offshore aquaculture during the 1990s. The Council of

Agriculture implemented a series of projects to enhance the capacity of Taiwan's offshore aquaculture. Not only did the Council promote offshore aquaculture, but it also facilitated the employment and improvement of a seawater supply system for inland aquaculture. This trend continued well into the fifth phase in the 2000s (Shi 2012: 125–126).

The fifth phase witnessed high aqua-cultural output, which amounted to more than 300 thousand tonnes per year between 2001 and 2009. As a result, the use of underground fresh water was also largely reduced from approximately 2.4 billion cubic metres per year in the early 1990s to 1.04 billion cubic metres per year in the fifth phase (Shi 2012: 126; 129). Nevertheless, in 2010 the largest form of aquaculture remained terrestrial: the area of seawater aquaculture was approximately 22 thousand square hectares; that of freshwater aquaculture was approximately 15 thousand square hectares; and that of offshore aquaculture was approximately 13 thousand square hectares. Moreover, the amount of production of inland aquaculture in the same year was roughly 27.4 thousand tonnes and that of offshore aquaculture reached 4.1 thousand tonnes (Shi 2012: 130).

Soon after the Sino-American Joint Commission on Rural Reconstruction (JCRR) moved to Taiwan in 1949, it began to help the ROC develop Taiwan's agriculture (including fishery). The director of the Fishery Division of the JCRR, CHEN Tongbai, was directly in charge of the development of aquaculture, demonstrating the director's high expectations of aquaculture. As the successive director of Chen, CHUEH Chuang-ti (2015), has identified, some important events took place in the development of aquaculture on modern Taiwan after 1949.

First, the cultivation and exportation of eels occurred in the mid-1960s. The highest amount of eel production was 80 thousand tonnes, and 80 per cent of the eels were exported to Japan (Chueh 2015: 62–63; 164–167). Second, the cultivation of giant tiger prawns was popularised in the early 1980s. In 1968, CHEN Tongbai invited aquacultural scientist LIAO Yijiu to come back to Taiwan and develop the technology for cultivating giant tiger prawns (Chueh 2015: 167–168), which grew rapidly between 1984 and 1987. The size of the cultivation increased 10 thousand hectares, and the amount of production increased 70 thousand tonnes within these four years (Shi 2012: 123). However, by the end of 1987, the aquaculture of prawns collapsed as a result of prawn disease (Chueh 2015: 171).

The third important event involved acquiring the technology to completely cultivate milkfish in 1979. Such an endeavour succeeded in 1982. To maximise production per hectare, CHUEH Chuang-ti advocated the method of deep-water cultivation using air pumps and artificial fish food. This method increased the production from two tonnes per hectare up to twenty tonnes per hectare (Chueh 2015: 173–174). The fourth event was the cultivation of mullets. In 1963, Taiwan's aquaculture specialists, TONG Yixiu and LIAO Yijiu, began to study the technology of cultivating mullets in 1963, and their success increased the amount of mullet production. Each year, more than one million mullets were produced, and the highest record of the production was four million (Chueh 2015: 174–175). The fifth development was the improvement of the technology of cultivating tilapia in the late 1960s and early 1970s. The quality of the tilapia produced with enhanced breeding and feeding methods increased tilapia exports to the US and Japan (Chueh 2015: 176–178).

The above events involved terrestrial aquaculture. Two additional events noted below were related to the use of sea aquaculture. The adaptation of offshore deep-water oyster farming in 1978 raised both the quality and quantity of oyster production. Additionally, small-scale offshore net cage aquaculture was initially conducted in several inland and offshore areas, for example Sun Moon Lake and Penghu in the 1970s. In 1993, the technology of deep-water near-shore net cage aquaculture was further explored to enlarge and commercialise this type of aquaculture (Chueh 2015: 182–183).

In 1996, the world's largest offshore net cage aquaculture was established by Asia Pacific Investment, a large commercial investment company in Taiwan. At the same time, the Pan Asia Ocean Business Corporation was also established to operate net cage aquaculture. However, three critical problems finally led to the termination of this corporation and its aquaculture business, as noted by Chueh (2015: 185–193), who was the then director of the governing board of the corporation.

First, there was no adequate regulation or proper provision of insurance for offshore deep-water net cage aquaculture. Second, no suitable vaccination could be utilised for the cultivated fish, resulting in a low rate of survival and high production costs. Third, undertaking offshore deep-water net cage aquaculture required permission by the state. Usufruct sea areas controlled and regulated by the ROC had to be approved by the concerned authorities. The state was reluctant to approve these rights for large corporations. Of the earlier three problems, Chueh (2015) emphasises the importance of fish vaccinations and medicine for fish diseases to reduce costs.

As the above introduction of the development and crucial events of Taiwan's aquacultural fishery shows, one can derive the following points regarding their constructions of the ocean through the lens of the Steinbergist Classification. The above introduction is to show how they used the ocean and what was important for them.

First, the sources of water for inland aquaculture were critical. People undertaking inland aquaculture on modern Taiwan often mix fresh water and salt water to lower the density of the salinity to increase production. Underground fresh water was excessively extracted for dilution as well as for use by freshwater aquaculture. Especially in the coastal areas, land subsidence and seawater intrusion were aggravated, and aquaculture fishponds could be easily damaged or even destroyed by the sea during torrential rains and typhoons. The sea, thus, became a source of threat.

Additionally, although the utilisation of more seawater for inland aquaculture was encouraged by the state and the policy was adhered to by the fishery (as demonstrated by the successful reduction in the amount of fresh water consumed for aquaculture mentioned earlier), seawater became a substitution for fresh water, but the ocean itself was still distant and not the focal point. Seawater reduced the consumption of underground fresh water. Paradoxically, seawater thus became a medium for inland aquaculture to ameliorate the threat of the intrusion of the sea and the sea *per se*. The sea, thus, was not viewed as a possessable resource reservoir. It was the water in the ocean that mattered for the security and even survival of inland aquaculture, that is, the seawater helped the industry defend itself against the threat of the sea.

Second, profitability was crucial. People undertaking and promoting the development of aquaculture on modern Taiwan strove to maximise the profits of this fishing industry. As shown in the earlier discussion of the phases and important events in the development of aquaculture, people leading the industry and state officials sought to improve and obtain the technology needed to cultivate more suitable fish and other aquaculture resources. Technological exploration, production orientation, disease prevention, market expansion, vending logistics, financial insurance, operational management all dovetailed for the purpose of increasing revenue and reducing costs. Modern Taiwan's aquaculture was, by and large, conducted by small-scale fish farmers (Chueh 2015) who were very sensitive to fluctuating prices of aquaculture produce. The ratio of cost-revenue ratio influenced the preferences and actions of those individuals. In effect, the importance of the above events identified by CHUEH Chuang-ti was underscored by the aquacultural trends in the choices of what was cultivated.

For example, in 1969, GUO He successfully interbred the tilapia of Mozambique with that of the Nile. The size of this new breed was larger and could grow faster than its predecessors. Moreover, in 1974, LIAO Yijiu managed to solve the problem of the overpopulation of tilapia larvae. Cultivating tilapia thus became more efficient and profitable. In 1979, tilapia production reached 34 thousand tonnes and tilapia became the most important freshwater aquaculture product in Taiwan (Shi 2012: 136). Another example can be found in the rise and fall of cultivating giant tiger prawns, which reached its apex in Taiwan in 1987, immediately before spherical baculoviruses struck a fatal blow to this shellfish. The value of the whole giant tiger prawn aquacultural industry was approximately 11.2 billion NTDs in 1987, but suddenly dropped to 470 million in 1988. The cost of cultivating giant tiger prawns became too high, and those who dedicated themselves to this type of aquaculture had to switch to another business (Shi 2012: 145). Above all, profitability was the main concern of the people engaged in aquaculture in Taiwan.

Third, the ROC's regulation and territorialisation of the sea affected the construction of the ocean by the offshore aquacultural industry in the 1990s. The aforementioned three obstacles identified by CHUEH Chuang-ti raised the cost of large-scale cultivation of cobias with offshore net cages. Although the shareholders of the Pan Asia Corporation did not want to continue their venture because of the low returns and high risks of offshore net cage aquaculture, small-scale individual operations continued this business in suitable sea areas, such as those of Penghu and Pingtung counties. Various fish of high economic values (for example, groupers and red seabreams) were raised by these small-scale aquaculture producers (Chueh 2015: 193).

Market demand for lucrative fish could still sustain small-scale offshore net cage aquaculture operated by individual producers because the opportunity cost was not a major factor impeding the continuation of their operations. Furthermore, after 2000, policies and technologies to solve many problems related to cultivating fish with net cages were developed and disseminated. The state played an important, supportive role in this development. For example, the state began to allow live groupers to be transported to Hong Kong after 2008 and the Chinese mainland after 2011 (Shi 2012: 141). An aquaculturist, WENG Pingsheng in Penghu, sponsored by the state, improved the material used to make the net cages for raising cobias in 2012. His invention largely increased the rate of cobia survival from less than 30 per cent to more than 60 per cent in 2013 (Zhang 2019).

As discussed in Chapter 2, after the 1990s, the ROC territorialised the surrounding sea areas of Taiwan (that is, twelve nautical miles of the territorial sea). The ROC also deemed it imperative to steward its fishing grounds in its EEZ. In terms of offshore aquaculture, the deployment of net cages in the sea and the 'fish farming' right *in rem* had to be granted and protected by the state (Chueh 2015: 194). More importantly, the way to raise fish using offshore aquaculture required the use of certain areas of the sea, and the net cages had to be fixed to the bottom of the sea. Such an aquaculture practice, combined with the notion of the right *in rem* and the compulsory application for state approval to use the sea areas for net cages and fish cultivation, as stipulated in the Fishery Act, constructed specific sea areas involved in fish farming, not the entire ocean nor the deep-sea, as exclusive, disposable and even possessable properties of the state by the offshore aquaculture of modern Taiwan.

Furthermore, in the case of Pan Asia Ocean Business Corporation, its largest investor, WANG Yongqing (also known as WANG Yung-ching, who was the wealthiest entrepreneur in Taiwan), did not maintain his financial support for net cage aquaculture because he could not secure the state's support for offshore aquaculture undertaken by large enterprises. In the mid-1990s (Chueh 2015: 189), Wang understood that the state did not wish to grant fishing rights to large corporations for the purpose of protecting the interests of small-scale aquaculture producers. However, by the early 2000s, the state itself started to promote small-scale offshore net cage aquaculture; at this point, fish farmers, as Chueh (2015) would call them, constructed the ocean in the style of the Seldenian Variation, as a result of their application for fishing rights *in rem*, of their interactions with the state regarding their offshore aquaculture activities, and of their utilisation of specific sea areas for their net cages and fish farming. This was a construction of the Seldenian Variation.

On the contrary, in light of the Steinbergist Classification, it can now be suggested that Taiwan's inland aquaculture constructed sector the ocean in accordance with the Indian Ocean Model. Not only did profitability become the primary concern of the people engaged in this type of aquaculture, but they also viewed the water in the ocean as an instrument to defend the erosion of land by the ocean. Even though they knew that what they were cultivating in their aquaculture ponds might originate from the ocean, the ocean was still marginalised in terms of its importance to their economic practices and material lives. This was a construction of the Indian Ocean Model.

Near-shore and Near-sea Fisheries

As the former president of the National Taiwan Ocean University, LEE Kuo-tien (2007) identifies, fishing activities within twelve nautical miles of the territorial sea can be categorised as near-shore fishery, whereas those within 200 nautical miles of the EEZs are classified as near-sea fishery. In this sub-section, I suggest that the near-shore fishery on modern Taiwan constructed the ocean in the style of the Indian Ocean Model; in turn, Taiwan's near-sea fishery constructed the ocean first following the same model and later absorbing elements of the Seldenian Variation.

The ROC's Fishery Act provided a legal basis to justify the state's stewardship over Taiwan society's fishing industries. However, the fishing industries' constructions of the ocean did not change as a result of the state's laws and ordinances. The aforementioned offshore aquaculture constructed the ocean according to the Seldenian Variation. This was not merely a result of the Fishery Act. The small-scale offshore aquaculture producers' practices and methods of using the ocean, combined with their reliance on the state's approval, support and promotion, combined to reflect construction of the ocean in the Mediterranean-Seldenian style.

In contrast, the near-shore fishery as well as inland aquaculture did not follow the state's construction of the ocean regarding fisheries. Even though they were requested by law to obtain the state's approval for their fishing activities, how they constructed the ocean, as shown in their actual economic practices, did not deviate from the Indian Ocean Model. Unlike in the case of offshore aquaculture, the state's policies and attitudes towards them did not influence their pre-existing constructions of the ocean. Thus, practices of ocean utilisation were critical as well.

Long before the Chinese people began to explore Taiwan and exploit its terrestrial natural resources in the Ming dynasty (1368–1644), the near-shore fishing industry around Taiwan Island had already emerged. The earliest record can be traced back to the Song dynasty (960–1279), when Chinese fishermen were chasing mullets in the sea areas between Fujian and southern Pingtung of Taiwan (Hu and Yu 2012a: 24). Understandably, the fishing techniques and boats used before industrialisation were relatively primitive. Both governments of the Ming and Qing (1644–1912) dynasties once prohibited civilian use of the ocean. Fishing activities were under severe legal restriction, and the fishing industries were considered marginalised economic activities (Hu and Yu 2012a).

Additionally, many people engaged in fishing in Taiwan also practised (and still do today) *wangye* (a god prince) worship. As a part of the worship practice, they would build and place daily necessities on a small boat and let it drift out to sea. Because they considered the ocean as a source of epidemics, they hoped such a practice would expel diseases and obtain the supernatural protection of *wangye* (Hu and Yu 2012a). The elementary school textbook containing a story describing a child worrying about the father going out fishing at sea in the dark night in severe weather conditions is widely known by the people of Taiwan. Although the textbook was used in the modern era, the traditional societal view about the dangers of fishing at sea was precisely summarised.

Such a representation of the ocean by the near-shore fishery did not change, and its conditions were not improved much when the Japanese colonial regime attempted to modernise Taiwan's fishing industries (Hu and Yu 2012b). Institutions and organisations of the production and distribution of Taiwan's fishery were established, for example, fishermen's associations. Many fishing ports were constructed, such as Donggang. Aquaculture and fishery research were promoted as well.

However, the Japanese colonial government aimed to use Taiwan's resources, including its fishery, to supplement Japan's own economy, as mentioned earlier. Therefore, the Japanese colonial administration took advantage of Taiwan's geographical location and used Taiwan as a steppingstone to exploit resources in the South China Sea and Southeast Asia. Taiwan's fisheries were then co-opted to meet such a goal. The relevant Japanese colonial bureaucracy on Taiwan concentrated its energy on developing the near-sea and deep-sea fishing industries (Hu and Yu 2012b).

Between 1895 and 1945, the near-shore fishery still suffered from its primitive fishing methods, equipment and vessels. In 1940, the production of near-shore fishery reached its apex before the end of the Second World War, but only amounted to approximately 23,000 tonnes, which was about 20 per cent of the total production of the fishing industries of Taiwan. However, soon after Japan surrendered and returned Taiwan to the ROC, the near-shore fishery continued to grow because Japan had confiscated larger near-sea and deep-sea fishing vessels during the War, and it did not require a large amount of money or highly industrialised technology and equipment to undertake near-shore fishing activities.

The near-shore fishery was a means for some people to earn their livings in the difficult post-war period. As an interviewee of this study, a retired captain of a merchant ship, relates, his father made an effort and asked a favour of a friend to become a near-shore fisherman to provide a living for the whole family after his father had brought his mother and brothers to Keelung Port on Taiwan from Shandong Province on the Chinese mainland in early 1950. His father had been an infantry-guerrilla officer before going to Taiwan. Catching fish (which seemed to be unlimited and abundant) on the sea (which seemed to belong to no one in terms of private property ownership) was viewed as a way to feed one's stomach and keep one warm.

In 1952, near-shore fishery production increased to approximately 44 thousand tonnes, roughly 36 per cent of the contemporary total fishery production of Taiwan (Ho 2012: 54–55). This trend was helped along by one of the state's economic policies in the early 1950s, which aimed for the near-shore fishery industry to absorb more unemployed labour, which continued until 1955. Thus, fishing harbours were repaired and built between 1945 and 1950. In 1953, unoccupied fishing boats were distributed to fishermen who needed vessels for a low price. After 1955, the state promoted near-sea and deep-sea fishery, after which the percentage of the near-shore fishery production of Taiwan's overall fishing industries began to decrease until the 1970s (Lee 2007: 67–68).

It is important to note that the state supported the near-shore fishery for at least ten years between 1945 and 1955, believing that it could provide more jobs in the fishing villages and even in the surrounding areas. The state also expected that the industry would provide more high-quality protein for its people (including soldiers) in Taiwan. The near-shore fishery was particularly important to the sustenance, and even prosperity, of small- to medium-sized fishing villages after the ROC withdrew to Taiwan.

Fishermen in this industry with less financial capital and technological equipment would set up homes in villages near their fishing grounds. When they returned to shore, their catches would be sold and distributed to the market. Certainly, some fish and other types of seafood in their catches would undergo processing before long-range transport and consumption. The logistics of the fish stocks and equipment (for example, net making, boat repairs and instruments and machine servicing) were able to sustain and develop the economy of fishing villages and, thus, contribute to societal stability.

This was what the state expected and what people tried to, and did, get from the near-shore fishing industry. In other words, the near-shore fishery became a safety net for the economy of rural-seashore areas and other fishing industries. Such a view and practice were well-reflected in the changing production of the near-shore fishery between the mid-1940s and mid-1980s. As mentioned above, production, which had reached its zenith in the mid-1950s, once again began to grow in the 1970s (when many states around the world proclaimed their EEZs and the oil crises hit the global economy), climbing to its peak of approximately 57 thousand tonnes in 1986 (Lee 2007: 68).

However, viewing and using the near-shore fishery as an economic safety net in difficult times by the state and people living along the shores of Taiwan did not result in the construction of the ocean as possessable property. For the state, that element originated from the international community, as discussed in Chapter 2. The people of Taiwan who were involved in near-shore fishing did not construct their fishing grounds as their private properties, but nor did they intend to protect or conserve the fish in the sea areas. The repercussion of overfishing, thus, worsened after the mid-1980s (Lee 2007).

In fact, the problem of overfishing had emerged earlier, in the late 1950s and early 1960s. As CHU Ming, a tycoon of the deep-sea fishing industry of modern Taiwan has noted, he was aware of such a problem before he established his fishery enterprise in 1961. Because of this near-shore overfishing problem, he was determined to embark on his deep-sea fishing venture (Chu 2010: 34). Furthermore, although the state made efforts to alleviate overfishing by prohibiting the use of gillnets, reducing and limiting the number of fishing vessels, setting up artificial reefs, conducting fingerling releases and promoting tourist-recreational fishing in the 1990s and 2000s, by the 2010s it had become necessary to decelerate the population loss in fishing villages (Chen, Wang-ching 2012; Chuang 2012; Ho 2012; Lee 2007). The population loss had resulted from the decline of fishery production, which was, in turn, a consequence of the problem of overfishing.

This chapter does not analyse how effective and efficient the ROC's policy of improving the living standards and economic conditions of near-shore and near-sea fishing industries and villages was. The above introduction of near-shore fishery aims to show how this industry on modern Taiwan constructed the ocean. As a reminder, in the late 1950s, the ROC promoted near-sea and deep-sea fishery. The percentage of near-shore fishery in the production of the overall fishery industry of Taiwan shrank. When the deep-sea and near-sea fishing industries faced difficulties, the near-shore fishery grew again.

However, the second decline happened in the late 1980s, when the fishing resources of the near-shore sea areas of Taiwan showed signs of depletion, which was followed by the decline of fishing villages. People started to catch fish in near-shore marine areas for their subsistence and because of the lower entry barriers to fishing. Their undertaking of near-shore fishing demonstrated no suggestion of constructing the fishing grounds as private properties. Fishermen caught fish and returned to their homes on land. When they could no longer catch sufficient fish to feed themselves, more and more fishermen gradually abandoned the near-shore fishery and left their fishing villages. This situation on modern Taiwan shows the construction of the ocean according to the Indian Ocean Model by the near-shore fishing industry.

When the tourist-recreational fishery in near seashore areas was promoted after the early 1990s, the ocean was constructed similarly as it is in the romantic oceanic literature. What mattered for the fishermen involved in this tourist industry and for the tourists utilising this kind of service was the scenery of the ocean. The ocean formed a background seascape. Seafood and the fishing experiences were services offered by the fishermen on the surface of near-shore sea areas. The quality of such service became the key to deciding whether the tourist-recreational fishery could be sustained and grows; instead of the ocean *per se*, what mattered were the living resources under the sea or adequate regulations on the use and possession of the ocean by the state. The construction of the ocean by the near-shore fishery (including its subtype, tourist-recreational fishery) could be classified as the Indian Ocean Model.

The near-sea fishery on modern Taiwan had embraced a similar construction of the ocean as that of the near-shore fishery until the 1990s. As mentioned above, the Japanese colonial government promoted Taiwan's near-sea fishery after it occupied this island in 1895 as part of Japan's efforts to exploit resources of the South China Sea and Southeast Asia (Hu and Yu 2012a; 2012b).

As Lee (2007) identifies, fishing within two hundred nautical miles of the EEZs can be classified as a near-sea fishery. Under Japanese domination, near-sea fishing boats of Taiwan would operate in the sea areas of East China Sea, east to the Zhejiang Province on the Chinese mainland, of the Daoyutai Islands and the Yaeyama Islands, and of the South China Sea near the Philippines and Sabah (Hu and Yu 2012b: 43, 47). As Hu (2012: 14) indicates, the near-sea fishery was revived in the mid-1950s. Trawl fishing on the Taiwan Strait prospered in the 1960s and 1970s. The growth rate of the production by the near-sea fishery industry was approximately 10 per cent per year before 1980. Its production in 1980 amounted to 370 thousand tonnes, approximately 40 per cent of the total production of Taiwan's overall fishery industry (Lee 2007: 69).

In terms of practices, the state had not yet paid sufficient attention to the problem of overfishing in the near-sea fishing grounds. As analysed in Chapter 2, it had to wait until the 1991 Fishery Act was stipulated, at which point the state absorbed elements of the Seldenian Variation from the international community. Before this time, the state's construction of the ocean regarding the fishery industry belonged to the Indian Ocean Model.

As did the near-shore fishery, the near-sea fishery constructed the sea consistent with the Indian Ocean Model. Indeed, after the mid-1950s, the state, through the JCRR, led the development of the near-sea fishing industry (Chueh 2015). The state could not reap the harvest of its policy without the fishermen's cooperation and support, however. For example, the fishermen were lukewarm towards the JCRR's promotion of pelagic pair trawling in the late 1960s and early 1970s (Chueh 2015: 106).

The state and the people engaged in near-sea fishing (as well as the deep-sea fishery, which I discuss later) solely aimed to increase production and maximise profits. Near-sea and deep-sea fishery did play a role in the state's plan to promote economic growth. As witnessed by CHUEH Chuang-ti (2015), the JCRR implemented many policies to facilitate the growth of the near-sea fishery industry. The policies were well-received by near-sea fishermen and made the construction of the ocean by the near-sea fishing industry consistent with that of the state at that time.

The first example was the promotion of new fishing equipment. The promotion of the use of nets made of synthetic fibre and nylon and of vessels with fossil fuel-driven engines in 1956 was the first state initiative for upgrading fishing technology (Chueh 2015: 103; Ou 2012: 69). The use of nylon nets and lines inevitably led to the employment of net and line haulers to prevent the occupational injuries of the fishermen on boats. Moreover, the utilisation of these machines could certainly increase fishing productivity (Chueh 2015: 101–104). Furthermore, with more advanced fishing equipment, it was necessary to find larger groups of fish to catch to maximise the efficiency of the equipment, as well as profits. Thus, electronic fishfinders were also promoted by the JCRR (Chueh 2015: 99–101).

Fishermen were initially suspicious about the usefulness of these instruments. However, their use was widely supported and accepted once the first few fishermen managed to profit from utilising the new equipment (Chueh 2015). Generally, obtaining and employing new production equipment required financial investment and fishermen's time and energy to master the operations' new procedures. This could increase costs and risk. Nonetheless, once there was a sign of profitability, fishermen were willing to take the investment risk and adopt the new way of fishing to achieve higher levels of production and more revenues.

The second example was the promotion of new fishing methods. Modern Taiwan's fishing industries had employed bottom trawling. Pelagic trawling was not widely utilised. The JCRR thus aimed to encourage this fishing method for the near-sea fishery in the early 1970s. The purpose of promoting pelagic trawling was not related to marine environmental protection and conservation. As Chueh (2015) identifies, a problem of pelagic trawling used by Taiwan's fishing industries in the 1970s and 1980s was that the size of the meshes on fishing nets was too small. The state did not regulate such an issue when pelagic trawling was promoted.

Moreover, when Chueh led the Fishery Division of the JCRR to promote this fishing method, he believed that the technique of pelagic trawling could help fishermen catch more fish as well as capture fast-swimming fish more easily. The JCRR also supported the attempt to improve this method, which succeeded in early 1975, and facilitated the revival of trawling fishing based in Keelung and Kaohsiung (Chueh 2015: 104–107).

The JCRR understood that Taiwan's near-sea fishery still employed ring netting. Ring netting was particularly popular after 1968 when fishfinders and net haulers became widely used. Most fishermen of the near-sea fishery based in Suao adopted this method. However, the ring netting fishery of Suao had no comparative advantage to Japan's purse seining fishery. Suao's ring netting fishery was particularly harmed by Japanese fishing activities east of Taiwan that employed the purse seining method in the early 1970s. In 1975, the JCRR began to promote the use of the Japanese purse seining method by the Suo fishery. Fishing vessels were upgraded, and the size of the vessels was enlarged from approximately 20 DWT to 200 DWT. The primary fish caught by Suao's near-sea fishery employing the new purse seining was mackerel. With JCRR's assistance, Suao's near-sea fishery was able to retain its competitiveness and continue to grow in the face of Japanese competition (Chueh 2015: 109–113).

The third example was the promotion of utilising the fish resources of bluefin tuna in the sea east of Taiwan. Taiwan's fishermen attempting to catch tuna had to sail to the South China Sea or the Indian Ocean. Bluefin tuna is a deep-water fish. Fishermen operating on the sea east of Taiwan did not realise there were bluefin tunas two to three hundred metres below the surface. They used to catch fish in the level between the surface and a depth of one hundred metres. After CHUEH Chuang-ti's promotion in the 1970s, the fishermen managed to master the skills and techniques for locating and catching bluefin tuna between April and July every year. The method of deep freezing to lower than minus 50 degrees Celsius was also introduced to maintain the quality of the harvested tuna and to enable their sale in the Japanese market (Chueh 2015: 120–122).

The promotion of catching bluefin tuna in the marine area east of Taiwan was not an easy task. Even DENG Huo-tu, the then director of the Fisheries Research Institute, considered the area 'a fishing desert' before the tunas were discovered (Chueh 2015: 121). Once valuable tunas were found and could be caught for profit, fishing operations were conducted in that area. Moreover, fishermen needed time to figure out how and when the tuna could be caught. Catching tuna became a hunting activity. Thus, the image of a fishing desert was turned into a fish-hunting field. This was still a construction of the ocean belonging to the Indian Ocean Model.

Nevertheless, different from the construction of the ocean by the near-shore fishery, the construction by the near-sea fishery changed after the 1990s. It can also be argued that for fishermen operating in the overlapping EEZs of Taiwan and the Philippines, the change occurred even earlier, in the 1980s. The construction was the Indian Ocean Model with the elements absorbed from the Seldenian Variation. The fishery could no longer focus on fish production and profit maximisation only. The state expected and requested, that fishing activities in disputed ocean zones be protected. It became inevitable that the ROC would have to handle this question to protect the interest of Taiwan's near-sea fishery. This was a result of the overlapping demarcations of the EEZs between Taiwan and the Chinese mainland, South Korea, Japan and the Philippines.

Amongst these clashing fishing interests in the overlapping EEZs, the most acute challenges were the cross-Strait fishing conflicts: the Taiwan-Japan and the Taiwan-Philippine fishing conflicts. Because Taiwan's near-sea fishing industry needed the state's prohibition (against the fishing activities by Chinese mainland vessels) and protection (from other state's harassment and assaults), modern Taiwan's near-sea fishery gradually absorbed elements of the Seldenian Variation of the Mediterranean Sea Model after the 1990s.

The United Nations Convention on the Law of the Sea (UNCLOS) was adopted and signed in 1982. During this time, the fishing capacities of the Chinese mainland, South Korea and Japan, and of Taiwan in the Yellow, East and South China Seas changed. The fishing capacities of Taiwan and Japan were diminishing, surpassed by South Korea in the 1980s, and the Chinese mainland took over South Korea in the early 1990s.

To ensure the fishing interests of their countries, Japan, South Korea and the PRC legislated their claims to their EEZs after the mid-1990s (that is, Japan in 1996, South Korea in 1996 and the PRC in 1998); they also entered into agreements regarding their overlapping fishing areas (Lee 2007: 71). Notably, the Philippines promulgated a presidential decree to claim its EEZ in 1978 and a National Marine Policy in 1994 as 'the guiding principles for management and development of the country's ocean space' (Batongbacal 2015: 416).

During these changing developments in marine-fishing in the region, the ROC that governed Taiwan and other surrounding islands was unable to participate in the arrangements as a result of the Chinese Nationalist-Communist Civil War (1946–1949). The ROC's absence from the agreements marginalised Taiwan from the new legal arrangements regarding the near-sea fishing industry in the disputed sea areas (Lee 2007).

In the case of the relationship between Taiwan and the Chinese mainland, fishing vessels of the latter often transgressed the so-called middle line of the Taiwan Strait. The middle line was a long-term result of political and military practices between Taiwan and the Chinese mainland, with no legal effect. Mainland fishing vessels often entered the sea area east of the middle line. Many of the vessels utilised the bottom trawling method with wheels attached to their nets, which severely damaged the marine environment of the fishing grounds and deteriorated the long-term fishing resources. Direct conflicts between mainland fishing vessels and those from Taiwan also occurred (Lee 2007).

Some fishermen from Taiwan loathed, and complained about, the mainland fishing activities that took place in their traditional fishing locations. Taiwan's near-sea fishery, thus, demanded that the ROC intervene and exercise stewardship over their fishing grounds, even though the politics between Taiwan and the Chinese mainland impeded the emergence of any fruitful solution (Lee 2007). The situation became worse when the cross-Strait relationship deteriorated, especially between the late 1990s and 2008. When the situation improved between 2008 and 2016, the problem was alleviated although not completely solved. Even the latest version of the ROC's ocean policy white paper published in 2020 still declares that the state intends to manage this problem (Council of Marine Affairs 2020).

In the case of the situation between Taiwan and its neighbouring countries, mainly Japan and the Philippines, the problem of conflicting fishing interests was not adequately solved until 2013 in the case of Japan and 2015 with the Philippines, under the MA Yingjiu (also known as MA Ying-jeou; hereafter MYJ) administration (Ma 2018).

On 9 May 2013, the Philippine coast guard fired their rifles at a fishing vessel with a ROC flag in the overlapping EEZs of the ROC and the Philippines, causing the death of the captain of the fishing vessel. The MYJ administration adopted a firm diplomatic stance over this incident. After the ROC's show of naval power and the MYJ administration's imposition of economic sanctions, the Philippines apologised and finally signed the Agreement on Law Enforcement Cooperation in Fisheries Matters between Taiwan and the Philippines, which provided a legal basis to defend against possible abuses of power by Filipino coast guards against Taiwan's fishermen operating in the overlapping EEZs (Ma 2018: 160–163).

Japan, anxious about Taiwan cooperating with the Chinese mainland on the issue of sovereignty over the Diaoyutai islands and unable to resist the political pressure of the US, finally conceded and signed the Taiwan-Japan Fisheries Agreement on 10 April 2013. This agreement established a marine zone covering approximately 74 thousand square kilometres, protecting the legal rights of Taiwan's fishermen to operate in this marine area while leaving the issue of sovereignty over the Diaoyutai islands unaffected (Ma 2018: 190–205).

Nevertheless, the following issue demonstrates that Taiwan's near-sea fishing industry did not construct the ocean entirely according to the Seldenian Variation, even though elements of this variation were assimilated by the fishery industry. This fishing branch did not coherently consider the sea areas of their fishing grounds as their property, even though the ROC had already claimed its EEZ and aimed to conserve the resources within these marine zones. Nevertheless, the near-sea fishing industry showed no intention of protecting or conserving marine resources in their fishing practices.

As mentioned earlier, UNCLOS was stipulated in 1982. In the 1990s, the Taiwan fishery lost access to the Yellow Sea, the Gulf of Tokin, the Gulf of Thailand, the coastal marine areas of Vietnam, the near-sea areas of Malaysia, the ocean areas north of Australia and the major portion of the East China Sea. It became too expensive or too competitive for Taiwan's fishery to continue their operations in those areas.

Almost all fishing vessels operating in the aforementioned areas returned to the ROC's EEZ around Taiwan. The consequences of years of overfishing gradually appeared after the 1980s. The number of mullets caught by Taiwan's near-sea and near-shore fishery decreased from approximately 2.53 million in 1980 to 218 thousand in 2003. Large yellow croaker production by trawling vessels was reduced from 707 tonnes per year in 1989 to 54 tonnes per year in 2002, and small yellow croaker production dropped from 3 thousand tonnes per year in 1989 to 146 tonnes per year in 2002. Anchovy production dwindled from 745.9 tonnes in 1989 to 38.3 tonnes in 2002. Moreover, the size of each type of fish caught was, on average, smaller than it had been prior to the 1980s (Lee 2007: 75).

Thus, the need for, and demands, of the near-sea fishery for state protection and prohibition in the conventional sea areas of its fishing grounds did not fully convert its construction of the ocean from the Indian Ocean Model into the Seldenian Variation. One can be certain that the near-sea fishery assimilated Mediterranean-Seldenian elements, especially after the 1990s, when the neighbouring countries of Taiwan Island established their EEZs that overlapped with that of the ROC. The absorption of the Seldenian elements originated from the requirement of state regulations. However, it was still necessary for the ROC to try to diminish the fishing capacity of the near-sea and near-shore fishery of Taiwan after 2016. The *2020 State Ocean Policy White Paper* continues to mention this goal and explicates the state's endeavours to establish the institution of responsible fishery (Council of Marine Affairs 2020).

No matter how much effort the state made, the unsatisfactory result of reducing the number of fishing vessels was closely related to the need by near-sea fishermen to earn their living from fishing. After other neighbouring countries institutionalised their EEZs, fishing activities in the near-shore areas of Taiwan also became an economic buffer for the people involved in this type of production. What the fishermen and the industry focused on in the practices of ocean utilisation was fish production and profit earning. Those were the two objectives that the near-sea and near-shore fishery strove to maximise.

To summarise, in this subsection I have examined the historical outline and important events in the development of modern Taiwan's near-shore and near-sea fishery. Their actual practices of marine utilisation, as well as a fisherman's poems (that is, a representation of the ocean) demonstrated their constructions of the ocean. The ocean was represented as the background of beautiful scenery for fishermen and people enjoying the service of the tourist-fishery industry. Furthermore, fishing on the sea was like hunting. Nevertheless, fish-hunting activities on the surface of the sea were also a way to acquire food and fibre in difficult times.

Naturally, new fishing methods, technology, equipment and vessels that could help maximise fishery production and generate more profits were encouraged by the state and welcomed by Taiwan's fishing industries. The conservation goals of the ocean's fishing resources were not taken seriously by the state until the early 1990s; neither was it a major concern for the near-sea and near-shore fishing industries. For fishermen in these industries, the marine areas of their fishing grounds were not considered anyone's private property and, to earn one's living, fishermen would make their greatest efforts using their best available technology to extract more fish or other aquatic species from the sea.

The ocean belonged to no one in the constructions by the near-sea and near-shore fishery, and fishing resource conservation was not their priority. At least, overfishing was not a concern of each individual fishery producer, even though the rational actions of each individual to catch more fish would ultimately lead to a self-defeating consequence for all of them, that is, the depletion of fish in their fishing areas. Thus, it can be argued that the constructions were formulated according to the Indian Ocean Model.

Although the near-sea fishery industry required the state's protection in the areas of Taiwan's overlapping EEZs with Japan and the Philippines and demanded that the state prohibit the transgressions by fishing vessels from the Chinese mainland after the 1990s, only then were the elements of the

Seldenian Variations added to their pre-existing construction of the ocean. I would not classify the fishery' construction of the ocean after the 1990s as a pure Seldenian Variation. Its pre-existing construction of the ocean, in accordance with the Indian Ocean Model, could still be demonstrated by the ROC's thus far unsuccessful attempt to reduce its fishing capacity. The question of fishing capacity reduction could also be found in the deep-sea fishing industry of modern Taiwan, which I discuss below.

Deep-sea Fishery

A former director of the Fisheries Agency of the Council of Agriculture, SHA Chih-i, indicates that deep-sea fishery on the high sea is a large-scale hunting industry (Sha 2007). Similar to the near-sea fishing industry of Taiwan during Japan's colonial occupation, this deep-sea hunting served the function of food provision for industrialised Japan. In the light of Taiwan's geographical location, its deep-sea fishery industry during this time utilised the fishing grounds in the areas of the Sulu Sea, Celebes Sea, South China Sea and the marine areas of Borneo (Hu and Yu 2012b).

Severely damaged during the Second World War, Taiwan's deep-sea fishery industry recovered very slowly in the decade immediately after the war ended. The difficulties faced by the deep-sea fishery industry were partially caused by the high entry barriers involving intensive capital and technological investments. These problems were aggravated by the lack of state support and subsidies at its nascent stage shortly after the war ended. As discussed earlier, the state expected that the near-shore and near-sea fishing industries would provide more job opportunities along the coastal regions of Taiwan Island. Before the mid-1950s, near-shore fishery played the main role in this regard, after which the state promoted the near-sea fishery. Nevertheless, the ROC provided few financial resources to assist the development of Taiwan's deep-sea fishery.

In 1961, the state offered loans to develop the deep-sea fishery and encouraged the construction of larger American-type purse seiners and steel tuna longliners. Taiwan's deep-sea trawling and tuna long lining ships gradually spread to the ocean areas near Australia, Africa and the Atlantic Ocean. In 1968, the state enacted a policy to actively promote deep-sea trawling and tuna long lining endeavours. Moreover, the deep-sea squid jigging and ultra-low temperature tuna lining industries developed at the same time. By 1973, the production of Taiwan's deep-sea fishery amounted to 362 thousand tonnes (Sha 2007: 37).

The deep-sea fishery, however, suffered a setback during the 1970s when the oil crises hit the world economy in 1974 and 1979 and when more than 40 states declared their 200 nautical miles of EEZs in 1977 and 1978. However, the growth of the deep-sea fishery was reinitiated in the 1980s (Sha 2007: 38). In 1983, a private company, Fong Kuo Fishery Group, adopted the purse seining method and built the first American-type purse seiner of one thousand DWTs for catching skipjacks and yellowfin tunas (Chueh 2015: 135).

In the same period, the exploration of fishing grounds for squid in the marine areas in the Atlantic Ocean near South America was successful (Chu 2010: 75). Moreover, the ultra-low temperature tuna lining fishery business became lucrative. Furthermore, in 1987, the state established the development of Taiwan's deep-sea fishery as the core of its fishery policy. The production of the fishery industry rapidly increased after 1983, reaching approximately 700 thousand tonnes in 1988 (Sha 2007: 38–39).

This growth trend continued well into the 1990s, although the ROC began to prohibit building additional fishing vessels under two thousand DWTs. New fishing vessels under two thousand DWTs could be constructed only as replacements after 1991. The state also prepared approximately three billion NTDs to purchase vessels aged fifteen years and above to accelerate the pace of replacement. Nonetheless, the deep-sea tuna long lining industry still managed to grow and produce 30 thousand tonnes per year in the early 1990s. The deep-sea squid jigging industry succeeded in developing the fishing ground in the sea areas near Peru in the 1990s, too. The total production of the industry in 1999 amounted to 276 thousand tonnes (Sha 2007: 39–40).

It is important to note that the deep-sea trawling industry had a different trajectory of development. Its conventional fishing grounds were generally located in other states' EEZs. Its production rapidly decreased in the 1990s and was only approximately ten per cent of the total production of Taiwan's deep-sea fishery (Sha 2007: 40). Using the fishing grounds in other states' EEZs required the payment of fishing access fees or other types of contributions and cooperation. Such requirements influenced not only the deep-sea trawling industry but also the development of the whole deep-sea fishery and its construction of the ocean after the late 1970s and early 1980s. I shall now discuss what typically constituted critical aspects of successful deep-sea fishing enterprises, largely according to the experiences of CHU Ming, a deep-sea fishery tycoon of modern Taiwan.

It can be argued that modern Taiwan's deep-sea fishery constructed the ocean in accordance with the Indian Ocean Model from the time of its inception in the 1960s to the period of the global trend of claiming EEZs by states around the world in the early 1980s. Its construction of the ocean can be demonstrated by the following important tasks related to preparing the operations of a deep-sea fishing corporation.

First, similar to the maritime shipping industries, suitable and seaworthy vessels were critical. Different from maritime transport services, the utilisation of fishing vessels was more complex. Each type of fishing vessel corresponded to a particular approach to fishing. Special fishing equipment was installed on specific kinds of vessels. Because of such complexity, running a deep-sea fishing business, as well as global maritime transport companies, required careful planning and calculations of costs and benefits, accurate market analysis and precise information and knowledge about fishing grounds.

Appropriate vessels utilising the appropriate fishing methods could then be identified and constructed (or purchased) according to the preceding evaluations. As mentioned earlier, CHU Ming had already realised the problem of the depletion of near-shore and near-sea fishing resources in the early 1960s. Therefore, when he first began to provide logistic services for Taiwan's fishery, he decided to offer the services for the deep-sea branch in 1961.

Additionally, when he actually embarked on his deep-sea fishing venture in 1965, he chose to undertake the tuna long lining fishery. He rationalised that the state had begun subsidising this deep-sea fishing branch and that this type of fishing had already existed on Taiwan for decades (Chu 2010: 35, 37). It would be less risky and more rational for Chu to start from this branch. Thus, he purchased three tuna longliners and caught fish in the marine areas of the Pacific Ocean. The first base of his fishing activity was located in Samoa.

Acquiring appropriate bases for deep-sea fishing activities was the second important task. CHU Ming gradually established additional bases in Cape Town, Mombasa, Singapore, Penang, Guam, Tahiti, Buenos Aires, Las Palmas Port, Abidjan and many other overseas ports. These bases made his deep-sea fishery economically viable. Because deep-sea fishing grounds are far away from Taiwan, returning to Taiwan from the fishing grounds to repair vessels and equipment, to resupply food and fuel and to unload and sell the fish required an unacceptable amount of time and money. The bases were generally chosen because they were close to the fishing grounds. Therefore, establishing overseas bases for logistic support of deep-sea fishing was an attempt to maximise profits and minimise costs (Chu 2010).

Nevertheless, as CHU Ming clarifies, the locations and conditions of deep-sea fishing grounds depended on the experience and judgement of the captains of the fishing vessels, although the fishing companies and concerned state bureaus also provided such information (Chu 2010; Chueh 2015). For any companies involved in deep-sea fishing, the third crucial task was to find appropriate captains to lead the ships.

A deep-sea fishing vessel worth ten million NTDs was considered a company, as Chu (2010) points out. An experienced and skilled captain was like a competent chief executive officer who could help the fishing corporation earn more profits. Moreover, when a captain led a fishing vessel, sailing and operating on faraway oceans, the captain's leadership, integrity and management had to be trustworthy. Therefore, in CHU Ming's case, he paid much attention to hiring suitable captains for his fishing ships.

Having caught abundant fish on deep-sea fishing grounds under the leadership of a competent captain, it was still necessary to sell the harvest to the market, which was the fourth key to successful deep-sea fishing ventures. One of the important criteria for choosing a suitable deep-sea fishing base was access to the market (including adequate processing and transportation facilities), even though the fish harvest might not be consumed locally in the regions of the location of the bases. Fish had to be sold to make profits for the fishing corporations.

Because fish caught in deep-sea fishery operations were often not consumed locally, fish processing became important for the targeted market. For example, tuna were usually sold to the US for canning before 1982 (Chu 2010: 68). However, the US tuna market was harmed by an economic downturn, after which the ultra-low temperature tuna long lining fishery expanded to supply tuna for Japan's sashimi demand. This also meant that the operations of transporting, freezing and refrigerating fish were critical for maintaining the freshness, quality and price of the fish.

Thus far, the above description of Taiwan's deep-sea fishery indicates that its construction of the ocean was aligned with the Indian Ocean Model. The practice of deep-sea fishery was not only like hunting on the ocean but also analogous to mining on the marine surface, especially when fish resources were still abundant. The ocean and its surface were not viewed as land. Fishing vessels were needed to mediate between fishing activities and the ocean. Fish were considered an unlimited resource hiding beneath the surface of the ocean, and extracting them required the appropriate equipment, methods and experiences and skills of ship captains and fishermen. The fish and other aquatic species harvested were like mined raw materials that had to be transported and processed for specific market demands. In this whole process, the ocean was viewed as merely a surface. No matter whether a sea area was a fishing ground or not, it was not viewed as someone's possession. This perspective was different from that of terrestrial mining but more similar to hunting in a wild field, representing a construction of the ocean that can be categorised into the Indian Ocean Model. This applied, however, before the 1980s.

After the early 1980s, the legal right to access certain fishing grounds became the fifth critical essence of Taiwan's successful deep-sea fishery industry. In 1973, 34 states declared their two hundred nautical miles of their EEZs (Huang and Lin, 2012: 84), and as mentioned earlier, this number increased to more than 40 by 1978 (Sha 2007: 38). After November 1994, when UNCLOS came into effect, this number grew to 164 (Huang and Lin 2012: 84). In effect, it became imperative for Taiwan's deep-sea fishing industry to secure its legal rights to fishing grounds.

There were three approaches to acquire these rights. First, bilateral interstate fishing agreements made between the ROC and the coastal states of the EEZs that still officially recognised the diplomatic status of the ROC could provide the legal bases for Taiwan's deep-sea fishing vessels to operate on the sea areas of the EEZs. However, the ROC's formal diplomatic ties were decreasing (Huang and Lin 2012: 90).

Second, civilian associations of fishery in Taiwan, that is, fishermen's associations, fishing company owners' unions and private fishing corporations (Huang and Lin 2012), could negotiate and sign fishing agreements with the coastal states of the EEZs. Examples of this type of deep-sea legal arrangement include the agreements between the Taiwan Tuna Purse Seiners Association and the Federated States of Micronesia (Sha 2007) and between the Taiwan Tuna Association and the Republic of Seychelles (Huang and Lin 2012).

Third, the fishing agreements of cooperation that could be sustained between Taiwan's civilian associations and their counterparts in the coastal countries of the EEZs were the most common of the three types of legal bases for undertaking fishing activities in the coastal states' EEZs. Most of modern Taiwan's deep-sea fishery cooperative arrangements existed in this form (Huan and Lin 2012; Sha 2007).

For these three types, fishing access fees were usually required. Sometimes, the coastal states demanded extra financial compensation or technological transfer. Some coastal states also stipulated a rule or law that requested that foreign fishing companies purchase what they harvested in the EEZs. The rule of law might stipulate that the aquatic species in the EEZs belonged to the citizens of the coastal states. Joint ventures and chartered arrangements were also common forms of fishery cooperation (Huang and Lin 2012; Sha 2007). The deep-sea fishing industry in the society of Taiwan managed to secure fishing access to the EEZs of approximately 30 states between 1999 and 2011. Every year, approximately seven to nine hundred Taiwan's deep-sea fishing vessels utilised any of the three forms of cooperation.

Based on this situation, it can be claimed that modern Taiwan's deep-sea fishing industry after the late 1970s and early 1980s absorbed elements of the Seldenian Variations from the international community.

As Steinberg (2001) convincingly suggests, the institution of the EEZ most resembles the Mediterranean-Seldenian construction of the ocean. 'The EEZ was established as a distinct zone between the twelve-[nautical] mile limit of territorial waters and the 200-[nautical] mile start of the high seas. In this zone, coastal states have exclusive rights to living and non-living resources, but for all other purposes high-seas freedoms prevail' (Steinberg 2001: 148–149). Moreover, the EEZs 'were conceived of as spaces for implementing a Seldenian variant of stewardship; EEZs were not so much to be developed as a component of state territory as they were to be managed as a space wherein a single state had exclusive rights' (Steinberg 2001: 171).

Therefore, it can be argued that those coastal states claiming their EEZs and allowing any form of fishery cooperation with Taiwan's deep-sea fishery industry absorbed elements of and constructed the ocean according to the Seldenian Variation. At the same time, when Taiwan's deep-sea fishing companies and groups submitted the requested fishing access fees, signed the agreements of cooperation, accepted the terms and conditions of fishing charters, or spent money on purchasing the fish harvested in the EEZs, Taiwan's deep-sea fishery acknowledged and recognised the rights granted to the coastal states by the EEZs. Consequently, the fishery's interactions with and the actual fishing practices on the marine surface carried out in compliance with the regulations and laws of the coastal states' EEZs, introduced elements of the Seldenian Variation into its pre-existing construction of the ocean.

However, regarding fishery, the international community did not stop at the creation of the EEZs. The question of fishery on the high seas was also regulated. According to the EEZ institution, the coastal state has the responsibility to conserve fishing resources within its EEZ. Fishing resources in the marine areas beyond or adjacent to the EEZs are then subject to the appropriate regional fishery management organisations. The purpose of the regulation is to conserve and develop this kind of fishing resource. Articles 63, 64 to 67, 87 and 116 address this issue (Huang and Lin 2012: 85).

The rights over the use and regulation of the straddling fish stocks, highly migratory species and anadromous species were involved in the disputes amongst states in the international community. Crucial conflicts between fishery interests included the US claim to its uninfringed right over salmon that spawned within its terrestrial territory and the US opposition to coastal states' regulation and management over the highly migratory species in their EEZs (Sha 2007).

The details of the related international politics are not discussed here. Nonetheless, the culmination of the politics was the 1993 United Nations Conference on Straddling Fish Stocks and Highly Migratory Fish Stocks. This conference 'adopted the Agreement for the Implementation of the Provisions of the 1982 United Nations Convention on the Law of the Sea Relating to the Conservation and Management of Straddling Fish Stocks and Highly Migratory Fish Stocks' (Steinberg 2001: 172). There were also other international treaties and regulations on the conservation of global fishery resources implemented in the 1990s, including the 1992 Convention on Biodiversity and the 1993 Agreement to Promote Compliance with International Conservation and Management Measures by Fishing Vessels on the High Seas. These conventions are hard law, with authority to legally constrain the actions of their subscribers.

The following are examples of soft law, which did not have legally binding obligations on states but instead demonstrated the prevailing norms and values in the international community that might become hard law in the future (Lin, Huang and Wang 2012: 250). Examples include 1992 Agenda 21, 1995 Jakarta Mandate on Marine and Coastal Biological Diversity, 1995 Code of Conduct for Responsible Fisheries, 1999 Rome Declaration on the Implementation of the Code of Conduct for Responsible Fishers and 2001 International Plan of Action to Prevent, Deter and Eliminate Illegal, Unreported and Unregulated (IUU) Fishing (Song 2004: 154–157).

The emergence and presence of these international treaties, agreements, codes of conduct and declarations on fishing activities on the high seas, in the form of both hard and soft law, not only illustrated the concurrent trend of norms and values but also revealed how the international community, by and large, constructed the ocean. No matter how constraining and binding the treaties and international agreements are, these hard and soft laws, in Steinberg's (2001) terms, can be viewed as the construction of the ocean in accordance with the Grotian Variation. This is because these international agreements or codes of conduct related to the high-sea fishery are Grotian instruments, and 'these Grotian instruments generally were implemented in a rather "weak" form' (Steinberg 2001: 125).

Three primary attributes of these hard and soft laws qualify their classification into the Grotian Variation. First, 'powers of stewardship often were given to non-state entities rather than states, under the assumption that they would be less likely than states to expand rights of stewardship to assertions of territorial enclosure' (Steinberg 2001: 125). Second, 'the signatory states claimed no right to control access to the sea or even behaviour in the sea but merely acceded to self-regulation to promote the long-term preservation of fish stocks' (Steinberg 2001: 129). Third, those agreements and treaties 'were merely statements by each state to abide by certain harvesting quotas, size restrictions, and/or equipment limitations within a bounded area of ocean-space' (Steinberg 2001: 128).

Above all, the ocean itself was still not possessable nor territorialised. The power of regulation was conferred on the community of concerned or participating major fishing states, but the necessity of stewardship over the limited fishery resources was acknowledged. The ocean was not external, irrelevant or ungovernable. This was a Mediterranean-Grotian style of construction of the ocean. The influence of the general construction of the ocean by the international community on Taiwan's deep-sea fishery industry could be seen in the gradual acceptance of the international norms and values emphasising the conservation and sustainable development of world fishery resources by the International Coalition of Fisheries Association (ICFA).

The ICFA, formed in 1988, was a fishery trade association representing the interests of major fishing countries' fish and seafood industries. Fisheries of modern Taiwan played an important role in the ICFA. The representatives of Taiwan's fishing industries could occupy critical positions in the ICFA. For example, CHU Ming had been the vice-chairman since its establishment until 1999. He was also elected as the chairman in 1994 and 1995 (Chu 2010: 115). It is worth noting that Chu (2010) has offered an account of the synopses of the annual conferences of the ICFA from the first meeting to that in the year 1999.

According to CHU Ming, in the first annual conference in 1988, the goals of preventing pollution of the ocean and protecting marine ecology were already addressed. In the annual conferences in the years of 1989, 1993 and 1994, speeches against unreasonable international regulations and restrictions, although unsupported by scientific evidence on deep-sea and high-sea fishery, could still be heard. Nevertheless, opposition to this position was expressed in speeches that recognised and supported the ultimate aim of protecting and conserving marine and fishery resources.

Starting in 1995, the attitude of the members of the ICFA shifted to favour the sustainable development of high-sea fishery. Discourse in the industry stopped attacking the general trend of the international community towards conservation and limiting the extraction of marine living resources. The ICFA began emphasising the value of reasonable fishing, especially its contribution to the production of food for the whole world. Since 1997, ICFA members have even started calling for the cessation of illegal and unregulated fishing and marine pollution (Chu 2010).

The changing attitudes of the members of the ICFA by no means infer that modern Taiwan's deep-sea fishery completely embraced the

Mediterranean-Grotian construction of the ocean. However, it can be suggested that the elements of this construction were assimilated. The ROC's stricter controls over the activities of its fishermen on the high sea after 2005 largely facilitated the absorption of elements of the Grotian Variation by Taiwan's deep-sea fishing industry.

As Lee (2007) points out and I as have mentioned earlier, the ROC had not given much attention to limiting Taiwan's deep-sea fishing capacity. The state had not made an attempt to solve the problem of overfishing caused by Taiwan's fishing vessels with the flag of convenience, either. Although the state had attempted to replace obsolete fishing vessels, serious efforts to reduce the number of vessels were made only under pressure by the International Commission for the Conservation of Atlantic Tunas (ICCAT) in 2005 (Lee 2007: 36).

Earlier in 2001, the Food and Agriculture Organization of the United Nations stipulated the International Plan of Action to Prevent, Deter and Eliminate Illegal, Unreported and Unregulated Fishing. Although it depended on the efforts of each concerned state and international fishing organisations' endeavours to implement the plan, many regional fishery management organisations began to request their member states to dispatch official vessels to prevent IUU fishing. Furthermore, all fishing vessels of the member states were required to provide detailed records of their harvests and to use vessel monitoring equipment. Additionally, statistical documents issued by flag states had to be delivered together with the fishing harvests. These methods and the norms and values behind them were incorporated by the ROC and Taiwan's deep-sea fishery after they were accused by Japan of engaging in IUU fishing in 2004 (Huang and Lin 2012: 87–88, 94–97).

In 2005, the ICCAT responded to Japan's request to drop Taiwan's tuna fishing quota to zero. Eventually, the ICCAT reduced the quota to 4.6 thousand tonnes for 2006 from 14.9 thousand tonnes in 2005. Only 15 of Taiwan's big-eye tuna-catching vessels were allowed to operate on the Atlantic Ocean in 2006. All vessels had to be boarded by ICCAT observers and to enter ports for inspection every three months (Huang and Lin 2012: 96).

Adequate actions were taken by the ROC, and Taiwan's deep-sea fishery cooperated with its state. Six billion NTDs were spent on dismantling 160 large tuna-catching ships and another 42 such vessels were ordered to return to ports and cease operations for one year. The methods widely used by the regional fishery organisations mentioned above were adopted and implemented. The ROC also promulgated the Act to Govern Investment in the Operation of Foreign Flag Fishing Vessels in 2009. These actions to regulate and manage tuna resource conservation not only allowed Taiwan's tuna fishing quota to be restored in 2006 and to be viewed as a positive model of conservation (Huang and Lin 2012: 96–97), but also demonstrates that, by the mid-2000s, modern Taiwan's deep-sea fishing industry generally integrated elements of the Grotian Variation.

A Brief Summary: Two Functions of One Story

The story provided above has two functions. First, important practices and experiences of the fishing industries in the society of modern Taiwan have been recounted while the constructions of the ocean by this marine-related economic sector have been interpreted. Details of the history do not need repetition here. However, it is still worth noting that the construction of the ocean by inland aquaculture can be classified as the Indian Ocean Model, whereas offshore aquaculture constructed the ocean in accordance with the Seldenian Variation. Additionally, modern Taiwan's near-shore and near-sea fishery both constructed the ocean consistently with the Indian Ocean Model. However, the latter assimilated elements of the Seldenian Variation after the 1990s. The Indian Ocean Model can also be applied to Taiwan' deep-sea fishing industry up until the late 1970s and early 1980s, when the fishery's construction of the ocean absorbed elements of the Seldenian Variation. Its construction also incorporated elements of the Grotian Variation in the mid-2000s.

The second purpose of the story in this section is to demonstrate one of the theses asserted by Steinberg's (2001) theory of territorial political economy. By and large, the narrative suggests that all branches of modern Taiwan's fishery initially constructed the ocean in accordance with the Indian Ocean Model, except small-scale offshore aquaculture supported by the state after the 2000s.

Although the deep-sea fishery industry later integrated elements of the Seldenian and Grotian Variations, and the near-sea fishery included elements of the Seldenian Variation, their original constructions of the ocean did influence the state's fishery policies and its intention to utilise the fishing industry for Taiwan's economic development in different periods. Moreover, and more importantly, their Indian-Ocean styled constructions of the ocean, together with the state's encouragement and utilisation of fishery, resulted in the excessive fishing capacity of Taiwan's fishery.

The problem of overfishing and fish stock depletion worsened once the ROC established its EEZ and had serious repercussions for the entire world. Over 90 per cent of larger fish resources in the high seas are estimated to have been overharvested in the past 60 years, with 32 per cent of the commercial marine species already facing the problem of extinction in the early 2010s (Huang and Lin 2012: 113). Therefore, the story's second function shows how the social construction of the ocean eventually affected both the material organisation of society and the ocean's first nature, as indicated by Steinberg's (2001) theory of territorial political economy.

Conclusion

In this chapter, I have presented two case studies. The first case study has offered a relatively simple thematic discussion of the development of important maritime transport corporations in modern Taiwan society, highlighting themes considered crucial by influential stakeholders in the maritime shipping industries. The second case study has offered a more complex historical narrative of Taiwan's fishing industries. I have interpreted the constructions of the ocean revealed by the above two case studies. In addition, I have used the second case as an example to demonstrate that the social construction of the ocean can influence the material organisation of society regarding the ocean and the geographical conditions of the ocean. This proposition is one of the theses pointed out by the theory of territorial political economy (Steinberg 2001).

In the first case study, the experiences of CMT, TN, YMMT (and its predecessor, CMSN), EM and WHL have been employed and elaborated. The critical factors (or themes) framing this case included ships, cargoes and terrestrial logistics support for the maritime shipping industry. Through evaluating these themes, it can be argued that modern Taiwan's maritime transport companies strove to increase their competitiveness and profits and to decrease the costs of their shipping services.

Possessing ships of sufficient quantity and adequate quality was crucial for the survival, continuation and development of a maritime transport corporation. Without the required ships, the corporation would be trapped in a financial abyss. Its development would be hindered, and even its survival threatened. Thus, it can also be argued that financial assistance was extremely critical for the maritime transport companies suffering and struggling to overcome such managerial and operational challenges.

Furthermore, owning ships was not the only requirement. Customising types of ships for shipping cargoes was also fundamental for the competitiveness and profitability of a maritime transport company. In the experiences of modern Taiwan's maritime shipping industries, containerisation was the key development and technological innovation that resulted in significant cost savings of maritime transport. Cost reduction by using containers to ship goods not only protected the goods but also minimised the time required for loading and discharging. Containerisation reduced cargo owners' costs and those of the maritime transport companies, thus increasing both profits and cargoes.

Maritime transport companies needed to transport the maximum number of cargoes, which was also closely related to the type and characteristics of ships. Attracting more cargoes required lower freightage and higher transport quality. The first factor mentioned above, deploying ships designed for specific cargoes, was a way to reach this goal; decreasing the costs of maritime transport reduced the price of shipping goods. Furthermore, undertaking and promoting the business of deep-sea liner shipping was another way to increase cargoes, especially when more products could be distributed overseas. To overcome the obstructions of freight conferences, helping cargo owners expand their market abroad increased the demand for maritime transport and the opportunities to ship more cargoes.

Nevertheless, creativity and mindfulness were also important when planning and arranging the details and operations of deep-sea liner shipping services, such as establishing overseas bases, deployment of ships and containers and port and cargo coordination. Sailing routes also needed thorough and careful consideration. A well-conceived route was highly cost effective, both reducing costs and generating more revenue. Moreover, forming alliances and implementing joint operations amongst maritime shipping corporations became a trend to cope with the requirement of using larger container carriers in a bustling, globalised era. The principles behind these approaches to acquire more cargoes were efficiency, cost reduction and higher profitability.

In terms of terrestrial logistic support, successful and influential maritime shipping entrepreneurs mentioned two critical issues for the development of maritime transport companies. Personnel recruitment and human resource management aimed at hiring the best talent and expertise for the maritime transport companies was the first critical logistics issue addressed by maritime shipping tycoons. A merit-based evaluation system for young graduates was promoted, and special internal training and education schemes were designed and implemented to consolidate and sustain smooth long-term business operations and development.

Logistical networks supporting the containerised maritime shipping industry around the world was the second critical issue. Building a network of liner shipping agents, establishing an inland container terminal and assembling a terrestrial container haulage fleet were keys for a large maritime transport company to increase the efficiency, quality and profits of its services. Furthermore, because the size of container carriers continued to increase, they entered fewer ports on different continents and regions, requiring near-sea maritime transport. Large deep-sea containerised liner shipping corporations cooperated with, or co-opted, near-sea shipping companies to complement their service provision.

Above all, modern Taiwan's maritime shipping industries aimed to maximise profits, minimise costs and maintain the competitiveness of their operations. These were the decisive factors that helped the classification of their construction of the ocean. Their construction of the ocean did not belong to the Micronesia Model. They were not attempting to possess part of the ocean nor generate wealth from the living or non-living ocean resources. Neither did they construct the ocean in the fashion of the Freitian Variation. Their experiences of providing services for the state and military were inconsistent with their efforts, as private companies, to develop and maximise profits. State intervention and obligatory services caused their financial deficits. Furthermore, in Steinberg's (2001) terms, modern Taiwan's shipping industries did not attempt to monopolise the entire channelled circulation of goods. Instead, they followed the spatial logic of terrestrial industrial capitalism, which strove to ensure smooth transport of goods across the surface of the ocean to lower the costs of production and distribution. The ocean was constructed as an empty surface, that is, the construction belonged to the Indian Ocean Model.

In the second case study, I have suggested that modern Taiwan's fishing industries constructed the ocean variously. Internally, each branch of the fishery (that is, near-shore, near-sea and deep-sea fishery and aquaculture) constructed the ocean in various ways. Externally, their constructions were different from that of the state, although the norms, laws and regulations of the international community influenced those of the near-sea and deep-sea fishery.

Developed and promoted mainly after the 2000s, offshore aquaculture, especially the small-scale offshore net cage aquaculture, constructed the ocean according to the Seldenian Variation. Its application for fishing rights, the acceptance of state guidance and support and the actual use of a portion of the sea for fish farming all support my interpretation and classification of its construction of the ocean.

For Taiwan's inland aquaculture, seawater, paradoxically, became an instrument to defend against erosion by the sea. The sea was a threat to the survival of inland aquaculture. Moreover, profitability continued to be the main concern of inland aquaculture. The ocean was still marginalised in terms of its importance to the practices and lives of the people undertaking inland aquaculture on modern Taiwan. Its construction of the ocean can be categorised into the Indian Ocean Model.

Similarly, modern Taiwan's near-shore fishery industry (including the tourist-recreational fishery) constructed the sea in the style of the Indian Ocean Model. The entry barrier for undertaking near-shore fishing in modern Taiwan society was low. Lower requirements for capital and technology allowed many people to join this economic sector, especially during difficult times. Moreover, the seemingly unlimited fish and unrestricted sea areas, together with the state's encouragement, became critical factors leading to the problem of overfishing and the depletion of fishing resources. Although the state did make efforts to alleviate the problem, as overfishing caused a decrease of fishery production, Taiwan's near-shore fishery lost fishermen and fishing villages suffered economic decline. Therefore, I have suggested that the construction of the ocean by the near-shore fishing industry was the Indian Ocean Model.

Prior to the 1990s, the near-sea fishery industry also constructed the ocean in the manner of the Indian Ocean Model. Modern Taiwan's near-sea fishery was promoted by the ROC in tandem with the deep-sea fishery. However, as mentioned above, the near-sea fishery still placed great emphasis on the maximisation of profits. Although the state encouraged the near-sea fishery industry to upgrade its fishing equipment and methods and to develop new fishing grounds, the fishermen of the near-sea fishery followed these recommendations only once they realised the profitability of the new fishing technologies.

Furthermore, the fishery did not view the ocean as a possessable reservoir of resources nor strive to conserve fishing resources. The ocean outside the fishing grounds was like a desert for the fishery industry. This construction of the ocean in the Indian Ocean Model absorbed elements of the Seldenian Variation largely after many states around the world established their EEZs.

Taiwan's near-sea fishery could no longer limit itself to production and profitability but began to request increasing state protection and intervention, both of which were key attributes of the Mediterranean-Seldenian construction of the ocean.

Notably, I have not argued that the near-sea fishery had converted its construction of the ocean completely to the Seldenian Variation, although some elements were assimilated due to the demands for state stewardship. However, the state's attempt to reduce overfishing was unsuccessful mainly because the fishery did not consider its fishing grounds to be personal property. Fishing resource conservation could, thus, not become the main goal of the fishery industry. In contrast, the primary goal was still for its fishermen to earn a living.

The construction of the ocean by Taiwan's deep-sea fishery initially reflected the Indian Ocean Model. Fishing was viewed as akin to large-scale hunting on the surface of the sea. Once the fishing grounds were discovered and developed, fishing became more analogous with mining on the ocean. During the 1960s and 1970s, deep-sea fishery, like the mining industry, focused more on factors contributing to the maximisation of profits and minimisation of costs, such as finding suitable captains for deep-sea fishing vessels, identifying valuable fishing grounds and establishing appropriate overseas bases for efficient production.

This construction following the Indian Ocean Model by the deep-sea fishery had absorbed elements of the Seldenian Variation by the 1980s. More and more deep-sea fishing companies made efforts to cooperate with coastal states and acquire legal access to their EEZs. When the ICCAT penalised Taiwan's deep-sea fishery and strongly demanded that the ROC solve the problem of overfishing, a series of actions to curb fishing capacity were undertaken. As demonstrated earlier, these actions and their consequences demonstrate that the deep-sea fishing industry on modern Taiwan incorporated elements of the Grotian Variation, especially by the mid-2000s.

Above all, apart from offshore aquaculture, all other branches of modern Taiwan's fishing industries, by and large, constructed the ocean consistent with the Indian Ocean Model from the end of the Second World War to when EEZs by coastal states around the world began to be established in the 1980s and 1990s. Taiwan's fishing industries used the ocean as a resource to make a living, according to the Indian Ocean Model. The fishery' operational rationale that aimed at maximising production and profits affected the ROC's fishery policies, resulting in excessive fishing capacity, which in turn led to declining fishery production in the state's EEZ.

The above historical experience demonstrates one aspect of Steinberg's (2001) theory of territorial political economy. That is, the social construction of the ocean can influence the material organisation of society and the geographical conditions of the ocean. The proposition demonstrated here consists of specialised terms, which have been explicated in Chapter 1. The other two main theses of the theory of territorial political economy have been elaborated in Chapters 2, 3 and 4. Having laid out the arguments of this chapter, I shall pull together all the threads of the main thesis of this monograph in the next, final chapter.

Note

1 This chapter is a modified and greatly expanded version of my journal paper, see Hou (2021).

References

Alphaliner (2011) 'Alphaliner Top 100: Operated Fleets as per 21 June 2011', *Alphaliner the Top 100*, accessible via https://public.alphaliner.com/.

Batongbacal, Jay L. (2015) 'The Philippine National Marine Policy', in Biliana Cicin-Sain, David L. VanderZwaag, and Miriam C. Balgos (eds.), *Routledge Handbook of National and Regional Ocean Policies* (Abingdon; Routledge), 416–443.

Chang, Yen-hsien (1997) 'Bianzhe Xu' (Editor's Foreword), in Yen-hsien Chang (ed.), *Zhongguo Haiyang Fazhan Shi Lunwen Ji, di liu ji*, Volume 6: Essays in Chinese Maritime History (Taipei; Research Center for Humanities and Social Sciences, Academia Sinica), I–III.

Chang, Yung-fa (1997) *CHANG Yung-fa Huiyilu: Changrong Jituan Zongcai yu Hai Jieyuan de Yisheng* (CHANG Yung-fa's Memoir: The President of the Evergreen Group and His Life Connecting with the Ocean) (Taipei; Yuan-Liou).

Chen, Bao-ji (2012) 'Xü Yi' (Preface One), in Dakuen Shen and Boyi Hong (eds.), *Gengyun Taiwan, Nongye Dashi Ji: Yuye Fenghua* (A Century of Innovative Agriculture in Taiwan: Fishing Prosperity) (Taipei; Council of Agriculture Executive Yuan, ROC), 4–5.

Chen, Kuo-tung (2013) Dongya Haiyu Yi Qian Nian, Er Ban (One Thousand Years of East Asian Ocean, Second Edition) (Taipei; Yuan-Liou Publishing).

Chen, Wang-ching (2012) 'Fuli Yucun, Fenghua Zaixian' (Rich and Beautiful Fishing Villages, Representing Their Prosperity), in Dakuen Shen and Boyi Hong (eds.), *Gengyun Taiwan, Nongye Dashi Ji: Yuye Fenghua* (A Century of Innovative Agriculture in Taiwan: Fishing Prosperity) (Taipei; Council of Agriculture Executive Yuan, ROC), 168–183.

Chi, Mengbin (2013) *CHI Mengbin Xiansheng Koushu Lishi* (The Oral History of Mr CHI Mengbin), recorded by Li Chang and Chin-lan Tseng (Beijing; Jiuzhou Press).

Chu, Ming (2010) *CHU Ming: Guoji Yuye Hezuo de Tuai Shuo*, Interviewed by Shenghuang Jian and Qiuming Lin (CHU Ming: An Advocate of International Co-operation of Fisheries) (Taipei; Academia Historica).

Chuang, Ching-ta (2012) 'Yuye Wenhua yu Yucun Xin Fengmao' (Fishery Culture and New Fishing Villages), in Dakuen Shen and Boyi Hong (eds.), Gengyun Taiwan, Nongye Dashi Ji: Yuye Fenghua (A Century of Innovative Agriculture in Taiwan: Fishing Prosperity) (Taipei; Council of Agriculture Executive Yuan, ROC), 184–209.

Chueh, Chuang-ti (2015) *CHUEH Chuang-ti: Taiwan Yuye Tuei Shou* (CHUEH Chuang-ti: Taiwan Fisheries Promotor) (Taipei; Academia Historica).

Council of Marine Affairs (2020) *2020 Guojia Haiyang Zhengce Baipishu* (2020 State Ocean Policy White Paper) (Kaohsiung; Council of Marine Affairs, Executive Yuan).

Goldstein, Joshua S. and Jon C. Pevehouse (2013) *International Relations*, 3rd edition (Boston; Pearson).

Ho, Chuan-hung (2012) 'Fen Jing He Ji: Cong Shanban Zhufa dao Gonghai Yuye Daguo' (Moving Forward Separately and Attack Together: From Sampans and Bamboo Rafts to Giant Fishing Country on the High Sea), in Dakuen Shen and Boyi Hong (eds.), *Gengyun Taiwan, Nongye Dashi Ji: Yuye Fenghua* (A Century of Innovative Agriculture in Taiwan: Fishing Prosperity) (Taipei; Council of Agriculture Executive Yuan, ROC), 54–65.

Hou, Kuang-hao (2021) 'Understanding the Social Constructions of the Ocean by Modern Taiwan's Fisheries', *Australian Journal of Maritime and Ocean Affairs*, DOI: 10.1080/18366503.2021.1933692

Hu, Hsing-hua (2007) 'Xulun' (Preface), in Shinwen Weng (ed.), *Taiwan Yuye Zhengce Yanjiu* (A Study of Taiwan's Fishery Policy) (Taipei; Chung Cheng Agriculture Science and Social Welfare Foundation), 6–9.

Hu, Hsing-hua (2012) 'Xie zai Lishi Changhe zhi Qian' (Before the Long Historical Record), in Dakuen Shen and Boyi Hong (eds.), *Gengyun Taiwan, Nongye Dashi Ji: Yuye Fenghua* (A Century of Innovative Agriculture in Taiwan: Fishing Prosperity) (Taipei; Council of Agriculture Executive Yuan, ROC), 8–21.

Hu, Hsing-hua and Mingcun Yu (2012a) 'Taohai Cangsang: Zaoqi Yuye Pian: 1895 Nian zhi Qian' (The Ebb and Flow of Fisheries: Early Period before 1895), in Dakuen Shen and Boyi Hong (eds.), *Gengyun Taiwan, Nongye Dashi Ji: Yuye Fenghua* (A Century of Innovative Agriculture in Taiwan: Fishing Prosperity) (Taipei; Council of Agriculture Executive Yuan, ROC), 22–35.

Hu, Hsing-hua and Mingcun Yu (2012b) 'Fengyun Jihui: Ri Zhi Shiqi Pian: 1895-1945' (In the Time of Transformation: Japanese Colonial Rule: 1895–1945), in Dakuen Shen and Boyi Hong (eds.), *Gengyun Taiwan, Nongye Dashi Ji: Yuye Fenghua* (A Century of Innovative Agriculture in Taiwan: Fishing Prosperity) (Taipei; Council of Agriculture Executive Yuan, ROC), 36–51.

Huang, Hsiang-wen and Lin, Ding-rong (2012) 'Yu Feng yingxiang Gonghai Yuye Da Guo: Yuanyang Yuye' (Riding the Wind to Become a Large High Sea Fishery Country), in Dakuen Shen and Boyi Hong (eds.), *Gengyun Taiwan, Nongye Dashi Ji: Yuye Fenghua* (A Century of Innovative Agriculture in Taiwan: Fishing Prosperity) (Taipei; Council of Agriculture Executive Yuan, ROC), 80–119.

Lee, Kuo-tien (2007) 'Taiwan Yan Ji Hai Yuye zhi Tantao' (Discussing Taiwan's Near-shore and Near-sea Fisheries), in Shinwen Weng (ed.), *Taiwan Yuye Zhengce Yanjiu* (A Study of Taiwan's Fishery Policy) (Taipei; Chung Cheng Agriculture Science and Social Welfare Foundation), 64–83.

Lee, Mee (1992) 'Zhonghua Minguo zai Taiwan Diqü zhi Haiyang Yunshu Fangzhan' (The Development of Maritime Transport of the ROC in the Area of Taiwan), in Yu-fang Lin (ed.), *Di Er Jie Ershi Yi Shiji Haichuan Yantaohui Lunwen Ji* (The Collection of Essays of the Conference on Sea Power in the 21st Century, the Second Round) (Taipei; Graduate Institute of International Affairs and Strategic Studies, Tamkang University), 233–253.

Li, Lianchi (1998) 'LI Lianchi Xiansheng Fangwen Jilu' (A Record of the Interview with Mr LI Lianchi), in Li Chang (ed.), *Haijun Renwu Fangwen Jilu, diyiji* (A Record of the Interviews with Naval Personnel) (Taipei; Institute of Modern History, Academia Sinica), 1–109.

Lin, Ding-rong, Hsiang-wen Huang, and Cheng-fang Wang (2012) 'Guoji Shiye ji Liangan Yuye' (International Perspectives and Cross-Strait Fisheries), in Dakuen Shen and Boyi Hong (eds.), *Gengyun Taiwan, Nongye Dashi Ji: Yuye Fenghua* (A Century of Innovative Agriculture in Taiwan: Fishing Prosperity) (Taipei; Council of Agriculture Executive Yuan, ROC), 249–267.

Ma, Ying-jeou (2018) *Ba Nian Zhizheng Huiyilu* (A Memoir of the Eight Years of Governing) (Taipei City; Global Views-Commonwealth Publishing Group).

Ou, Ching-hsiewn (2012) 'Zuoguo Jieduan Renwu, Chidai Shoushen: Jinhai Yuye' (Temporary Task Accomplished, Expecting Slimming Down: Near-sea Fishery), in Dakuen Shen and Boyi Hong (eds.), *Gengyun Taiwan, Nongye Dashi Ji: Yuye Fenghua* (A Century of Innovative Agriculture in Taiwan: Fishing Prosperity) (Taipei; Council of Agriculture Executive Yuan, ROC), 66–79.

Sha, Chih-i (2007) 'Taiwan Yueyangyang Yuye Huigu yu Zhangwang' (Retrospect and Prospect: The Deep-sea Fishery of Taiwan), in Shinwen Weng (ed.), *Taiwan Yuye Zhengce Yanjiu* (A Study of Taiwan's Fishery Policy) (Taipei; Chung Cheng Agriculture Science and Social Welfare Foundation), 34–63.

Shi, Shenglong (2012) 'Di Yi Zhang, Di Si Jie: Buduan Tupuo Pingjing, Chuangxin Jiazhi: Yangzhi Yuye' (Continuous Breakthrough, Creating Values through Innovations: Aquaculture), in Dakuen Shen and Boyi Hong (eds.), *Gengyun Taiwan, Nongye Dashi Ji: Yuye Fenghua* (A Century of Innovative Agriculture in Taiwan: Fishing Prosperity) (Taipei; Council of Agriculture Executive Yuan, ROC), 120–151.

Song, Yann-huei (2004) 'Guoji Yuye Guanli Fagui Xin Fanzhan Qushi' (New Developments in International Fisheries Management Laws and Regulations), in Weicheng Su and Yaojie Zhou (eds.), *Fisheries Research Institute Special Publication Number 5: Fisheries Bycatch and Discard Problems in Taiwan* (Keelung; Fisheries Research Institute, Council of Agriculture), 151–161.

Steinberg, Philip E. (2001) *The Social Construction of the Ocean* (Cambridge; Cambridge University Press).

Su, Maosen (2007) 'Yongxü Shuichan Yangzhi' (Sustainable Aquaculture Fisheries), in Shinwen Weng (ed.), Taiwan Yuye Zhengce Yanjiu (A Study of Taiwan's Fishery Policy) (Taipei; Chung Cheng Agriculture Science and Social Welfare Foundation), 84–107.

Tai, Pao-tsum (2000) *Jindai Taiwan Haiyun Fangzhan* (The Development of Maritime Transport on Modern Taiwan) (Taipei; Yushan She).

Wade, Robert (2004) *Governing the Market: Economic Theory and the Role of Government in East Asian Industrialization*, 2nd edition (Princeton; Princeton University Press).

Wang, Yu-fong (2016) *Puolan Zhuangkuo: Taiwan Huogui Zhuanyun Shi* (Magnificent Wave: Taiwan's History of Container Transport) (Taipei; Commonwealth Publishing).

Zhang, Zhihe (2019) 'Penghu Xiangwang Yangzhi Youcheng, Nian Chanzhi Puo San Yi' (Success in Net Cage Aquaculture in Penghu, Production Worth of More than 300 Million NT Dollars Per Year), *Penghu Times*, published on 14 November 2019.

6 Conclusion

A Foundation for the Political Analysis of Marine Policy[1]

Introduction: Elucidation and Recapitulation

In this concluding chapter, I summarise what I have discussed thus far in the monograph and attempt to formulate an analytical framework for future studies of the politics of marine policies. In the previous chapters, I have presented historical and interpretive accounts of how the sources of social power in the society of modern Taiwan constructed the ocean between 1949 and 2016. As suggested in Chapter 1, this academic pursuit in the form of interpreting the social constructions of the ocean can offer a foundation for understanding the politics of stipulating and implementing the state's marine policies. The constructions of the ocean can be considered the social structures in which a person or group takes political actions in the processes of marine policies. Therefore, a summary of the constructions elaborated in the previous chapters will be instrumental to the formulation of the framework for the political analyses of marine policies that will be presented later. The summary and the framework are the two main parts of this chapter.

Clarification of Concepts

The introductory paragraph uses the term 'social structure', which requires a brief clarification. A structure can be defined as a complex phenomenon or entity, which consists of its components, the relationships amongst the components and the patterns of interactions of the components. A social structure can be defined as a complex phenomenon comprising people (or organisations), their relationships with each other and the patterns of their interactions. For example, the students at a small university department can have their own social structure, which comprises the individual students, their subgroups and the normal interactions of the students within and between these subgroups.

Another relevant point worth clarifying is that the term 'social' is viewed differently from 'societal' in this monograph. This different treatment of these two words is based on Michael Mann's (2012a) definition of society. The word 'social' simply refers to human-to-human relationships. A group

DOI: 10.4324/9781003291343-6

of any number of people, from two or more, can establish a social relationship among its members. However, the word 'societal' is more complex and purely of society. It is more specific and refers to what belongs (and is subject) to society. For example, the term 'social power' means the power of and between individuals, groups and organisations, whereas the term 'societal powerholders' denotes powerful actors of a given society (if it were effectively possible to delimit the scope of a society). The earlier distinction between 'the social' and 'the societal' can be better understood when I utilise Mann's (2012a) definition of society below. The distinction helps readers comprehend the main theme of this monograph: The social constructions of the ocean in the society of modern Taiwan.

It might be abstruse to elaborate on what I mean by 'structure', 'the social' and 'the societal' in the introductory section of the final chapter of the monograph. It might also be considered unnecessary to summarise the primary analytical frameworks and their principal concepts employed in this study. However, conceptual clarification not only reminds readers what those abstract terms of previous chapters refer to but also facilitates, for those who have skipped other chapters, an understanding (if not appreciation) of what and how I have argued and interpreted in this book with greater ease.

The Precis of the IMES Model

As elaborated in Chapter 1, Mann's (2012a) historical-political sociology starts from his notion of society, which is defined as overlapping and intersecting networks and institutions of social power. According to this notion, these networks are actually sources of social power. His concepts and theories culminate in his IEMP Model (or IMES Model as used in this monograph), comprising at least three main elements (Hou 2012).

First, he convincingly suggests that the most important sources of social power can be classified into four categories: Ideological, economic, military and political (hence, the IEMP Model). As I have clarified in Chapter 1, because Mann's (2012a; Hou 2012) 'political sources of power' refer to the power of the state, I have changed this term to 'state sources of power' (hence the IMES Model).

Second, in the IMES Model, macro social changes can be explicated and comprehended by the four irreducible and inseparable categories of the sources of social power. The IMES Model identifies that a coordinating political, organisational centre can establish a particular network of social power, drawing institutions and resources from other types of sources of power in various proportions to satisfy human needs. Because human desires are most likely limited only by people's own imaginations, any existing coordinated networks of power cannot completely solve every problem, placate all the discontented or cage all sources of power. That is, problems, discontents and dissatisfaction can lead to the emergence of social interstices among different networks of power, where individuals, groups and organisations whose needs are ignored or unsatisfied may be abandoned. Once a political organisation begins to coordinate these individuals, groups or organisations, a new network of social power is formed, which is highly likely to challenge or clash with the pre-existing, dominant network of power, causing macro social change (Hou 2012; Mann 2012a).

This second element means that historical-political sociologists need to explicate and interpret social changes at the comparatively macro levels by considering the four types of the sources of social power as a comprehensive set without invoking a final, single factor as the ultimate cause (Hou 2012; Mann 2012a). That is, social change should not be deemed ontologically reducible and analytical reductionism should be minimised.

Third, the IMES Model also deals with the structure-agency polemic. The IMES Model does include agency in the explication and interpretation of social change. Agency plays a critical function in the analyses of social change according to the IMES Model. Leaders, elites and powerholders usually take actions to satisfy human needs, solve problems and overcome challenges but, at the same time, result in unintended consequences (Hou 2012; Mann 2012b). This view is endorsed and employed by this monograph. The analytical framework proposed later for understanding the politics of marine policies follows the same stance as the IMES Model regarding structure-agency polemics.

The stance of the IMES Model can be succinctly summarised with a direct quote from Karl Marx (1996: 32): 'Men make their own history, but they do not make it just as they please in circumstances they choose for themselves; rather they make it in present circumstances, given and inherent'. In the IMES Model, state leaders

> do not make decisions in a vacuum isolated from the other [societal] [powerholders]. Each of these [powerholders] takes decisions and actions that influence all others in an unintentional way. Thus, no [powerholder] can control the whole situation.
>
> (Hou 2012: 196)

This third main element of the IMES Model is important for demonstrating the utility of the framework that I attempt to build later in this chapter.

As mentioned in Chapter 1, the first element is utilised in the previous chapters of this monograph, whereas the second and third are not. The third element, however, is particularly important for this final chapter because the framework for the politics of marine policies presented later places more emphasis on actors and agency than on institutions and structure. The third element of the IMES Model provides a theoretical counterweight to balance the stress. Thus, this element connects the interpretations of the social constructions of the ocean in the previous chapters and links the constructions to the framework offered later and any other future studies employing the framework. In other words, the social constructions discussed in the previous chapters become the structure (and basic context) against which the framework presented later can examine the actors. The third element of the IMES Model bridges the framework and the structure in which powerholders in the society of modern Taiwan undertook actions during the design and implementation of marine policies.

Having clarified the distinction between the terms 'social' and 'societal', identified the meaning of 'social structure' and explicated the function of the analyses of the social constructions of the ocean in the previous chapters, it is sensible to remind readers of some of the critical concepts used in this monograph. The concepts that are briefly repeated here, and which can be found in Chapter 1, are politics, resources, institutions and power.

The Synopsis of Politics, Resources, Institutions and Power

Throughout this monograph, the term 'politics' follows Adrian Leftwich's (2004b) definition. Politics comprises all the interactions, including negotiations, compromises, acts of co-operation and conflicts, between and amongst individuals, groups and organisations, and the power exercised in these interactions to organise, and often institutionalise, the production, distribution and use of human and natural resources. Politics thus defined refers to the processes of human and institutional interactions that occur in our daily lives. It is more than activities that take place only within certain sites, such as between the executive and legislative branches of a government. Thus, how the state develops and implements its marine policies within its territory, or the parameters of its power is a political process.

Developing a social construction of the ocean is politics as well. As Philip E. Steinberg (2001) points out, the social construction of a geographical phenomenon (or any part of nature), for example the ocean, consists of its use, regulation and representation. The process of using, regulating and representing the phenomenon not only involve the interactions between people and nature but also, inevitably, among individuals, groups and organisations. This view also corresponds with Steinberg's (2001) theory of political economy, which convincingly suggests that the social construction of nature, the material organisation of society and nature *per se* are mutually influenced.

Notably, as shown above, politics in Leftwich's (2004b) definition is closely related to resources and institutions. First, as elaborated in Chapter 1, resources refer to material and nonmaterial entities that humans use to achieve their goals individually or collectively (Leftwich 2004b). Scarcity and the uneven distribution of resources can result in asymmetrical structures between the participants who engage in interactions, although the possibility of co-operation is not necessarily diminished by such asymmetry. Second, as also explicated in Chapter 1, institutions are required if the participants of interactions intend to cooperate for a longer time and to obtain goals effectively and even efficiently. An institution in this monograph denotes a set of patterned interactions, which can be formal or informal rules structuring actions. Moreover, the patterned interactions must last for a certain period of time and sustain a certain degree of stability. Members of the set of patterned interactions share some common norms and values regarding the institution (Peters 1999).

Resources and institutions are also two critical essences of power. Anthony Giddens (1987) and Mann (2012a) both define power as transformative capacity, which can intervene and alter a given set of events and conditions (Giddens 1987: 7). Power as transformative capacity is exercised through resources and institutions. As discussed in Chapter 1, unevenly distributed resources and unequally commissioned tasks (and assigned positions) in an institution can generate an asymmetrical power relationship between actors. However, through deliberate political actions and institutions, asymmetrical power relationships are not necessarily destined to and zero-sum dynamics; negotiations, compromises and co-operation are possible, and collective power (and even synergy) can be obtained (Hou 2012; Leftwich 2004b; Mann 2012a).

The Résumé of the Steinbergist Classification

Readers may find it convenient if I repeat here the core of the Steinbergist Classification. The Steinbergist Classification, as mentioned in Chapter 1, is shorthand for 'the classification of the social constructions of the ocean'. As identified by Steinberg (2001), the Steinbergist Classification consists of three primary models: The Indian Ocean Model, the Micronesia Model and the Mediterranean Sea Model. The core of each model is summarised later.

The Indian Ocean Model socially constructs the ocean as a void or an empty space. The ocean as empty space can be neither possessed nor governed, but trade and migration can be undertaken across it. Consequently, in this model, the ocean is still marginal to society. Wealth and higher social statuses are acquired primarily along inland rivers with convenient access to large agricultural surpluses in pre-modern, traditional civilisations. In modern, industrialised civilisations, the focus is on territorial investment and disinvestment for capital reproduction with little, if any, regard for the ocean (Steinberg 2001). In extreme cases, the ocean can become a source of threats in the Indian Ocean Model.

In the Micronesia Model, the ocean is constructed as a property and resource repertoire, which can be territorialised by states. The ocean provides daily necessities for both pre-modern and modern societies in the Micronesia Model. Moreover, the resources that the ocean offers include 'connections', as well. The seas are considered highways places in their own right that connect disparate lands instead of spatial barriers that must be crossed and overcome, creating hindrances to the travel of people and goods. In contrast, the ocean facilitates travelling and transport. The ocean as a place is viewed as the centre rather than the periphery in the Micronesia Model (Steinberg 2001).

The third model of the Steinbergist Classification, the Mediterranean Sea Model, comprises three variations. In the original version, as in the Indian

Ocean Model, the ocean must be dominated militarily by states to ensure the safety of the transport of military personnel and equipment. Each of the variations—the Freitian, the Grotian and the Seldenian—present particular modifications to the standard model.

The Freitian Variation is similar to the original Mediterranean Sea Model in that it also considers the seas as an object of control by states. Unlike the original model, which emphasises military domination for security reasons, the Freitian Variation places more stress on state authority for trade. According to the Freitian Variation, the state seeks not only to maintain safety but also to facilitate trade along the sea routes. The Grotian Variation does not view the ocean as possessable nor marginal to society. This variation also deems it necessary for the ocean to be governed, but by the global community of states, not by a single state. No individual or organisation should interfere with the access of others to the ocean for marine navigation and fishing. The community of all states, conferred with the right to govern the ocean, should allow and protect free access for all to the seas. The Seldenian Variation sees the sea areas of fishing grounds and the coastal marine zones as resource providers for states and, thus, territorialisable. Unlike the Micronesia Model, which views the deep ocean as a void that cannot be owned and provides no resources, the Seldenian Variation considers that sea routes for trade can be governed by states even while ensuring and protecting freedom of navigation for all (Steinberg 2001).

Having summarised the Steinbergist Classification, the chapter continues with a discussion of the three main theses of the theory of territorial political economy (Steinberg 2001). These arguments can also be better understood when the social constructions of the ocean by the four most important types of social power in the society of modern Taiwan are summarised in the next section.

The Social Constructions of the Ocean in the Society of Modern Taiwan: A Summary

Chapters 2, 3, 4 and 5 of this monograph interpreted the social constructions of the ocean by the four most critical sources of social power: The state, ideology, the military and the economy. This interpretation is conducted through the lens of the Steinbergist Classification. Notably, the classification is based on Steinberg's (2001) theory of territorial political economy, which comprises three main theses.

First, Steinberg's (2001) theory of territorial political economy identifies that the social construction of nature consists of the use, regulation and representation of nature. Its use, regulation and representation are mutually influenced. Changes in each element affect the overall social construction of nature. Chapters 2 and 3 not only explicate the constructions of the ocean by the ROC and the ideological power sources of Taiwan but also exemplify and illuminate this main thesis of Steinberg's (2001) theory.

Second, nature influences the material organisation of society and, in turn, influences the social construction of space. Chapter 4 attempts to demonstrate this thesis in addition to interpreting the constructions of the ocean by the ROC's Armed Forces. Third, the social construction of space influences the material organisation of society and, in turn, influences nature. Chapter 5 endeavours to illustrate this thesis as well as construing the constructions of the ocean by Taiwan's maritime shipping industries and fishing sectors.

Several of the essential concepts of Steinberg's (2001) theory, presented in my introduction in Chapter 1, are briefly elaborated again here. The term 'nature' refers to geographical phenomena, such as mountain ranges and river systems. The term 'material organisation of society' is a set of institutions and technologies. The term denotes the institutions and technologies utilised to transform and exploit nature, for example, the institutions of industrial capitalism and the related means of production. The 'social construction of space' refers to the uses, regulations and representations of certain geographical phenomena. For instance, the people of southern Fujian Province considered the seas as an immense and dangerous space across which a wealthy, happy migratory life can be pursued and achieved.

The Social Construction of the Ocean by the State Governing the Society of Modern Taiwan

As mentioned in Chapter 2, the state is a member of the society located in its territory (Migdal 2001). The state, for Mann (2012a), is the major political source of social power. The state can be defined as 'a political organisation embodying centrality, in that political relations radiate to and from the centre to cover a territorially demarcated area over which this political organisation exercises a certain level of authoritative rule and which is supported by organised military forces' (Hou 2014: 191; Mann 2012a).

The ROC has been the most important political organisation (effectively governing Taiwan and its affiliated islands, Jinmen and Mazu) in the society of modern Taiwan since 1949. Prior to 2016, the ROC constructed the ocean based on its economic policies, and after 2000, according to the Democratic Progressive Party's (DPP's) efforts to cultivate a national identity supporting the Taiwan Independence Movement (TIM). The state's construction of the ocean between 1949 and 2016 can be examined through its maritime and marine laws, as well as ocean policy white papers, which show how the state represented, regulated and used the ocean. By investigating how the state constructed the ocean, it is possible to demonstrate that the state's use (and intended utilisation) of the ocean influenced its regulations and representations of the ocean, thus illustrating the first main thesis in Steinberg's (2001) theory of territorial political economy.

The ROC's army was defeated by the People's Liberation Army (PLA) led by the Chinese Communist Party (CCP) in the Nationalist-Communist Civil War (1946–1949), after which the state capital was relocated to Taipei.

By and large, the state's construction of the ocean between 1928 and 1949 (when the Chinese mainland was nominally dominated by the Chinese Nationalist Party, or Kuomintang; KMT) can still be classified as the Indian Ocean Model. This is because the ROC, led by the KMT, still focused on dealing with constant terrestrial warfare on the mainland. The affairs of naval forces, marine merchants and fishing fleets attracted little attention from state leaders, although the state did stipulate the Maritime and Fishery Acts in 1929. These acts (expressing the ROC's will and proclamations regarding the governance and management of the relevant challenges and problems in maritime and fishery affairs) did not contribute to any significant changes in the state's overall attitude towards and construction of the ocean.

However, changes occurred after the state moved its operations to Taiwan at the end of 1949. The ROC, governing and maintaining Taiwan as the ultimate bastion for the expected counterattack against the CCP and the PLA, understood that it was also imperative to promote Taiwan's economic recovery, growth and industrialisation. Creating a sound industrial capitalist economy became one of the state's primary objectives (Wade 2004).

The Maritime Act was overhauled in 1958. The state aimed to improve Taiwan's maritime shipping industries and to build a solid legal framework for maritime transport services. The state needed these services to facilitate the economic policies of, first, import substitution and, later, export orientation. Maritime shipping industries were critical for imports and exports. Furthermore, the ROC enhanced its governing of Taiwan's international trade and maritime shipping industries by legislation that created the 1981 Shipping Act, which fortified Taiwan's shipping system and strengthened its shipping industries and economy. According to the Shipping Act, the state implemented policies to develop its shipping industries, continuing its adjustment and co-operation with industry.

As discussed in Chapter 2, through evaluating and interpreting the 1958 Maritime Act, the 1962 Maritime Act and the 1981 Shipping Act, it can be argued that the ROC incorporated elements of the Freitian Variation into its social construction of the ocean. The ocean itself did not occupy the central position in the state's economic development. Instead, the state intended to govern and encourage the use of the marine surface to promote international trade through the establishment of a suitable legal framework.

Before the end of the 1990s, the ROC's construction of the ocean had assimilated elements of the Freitian Variation, as demonstrated in its related maritime laws. As I have suggested in Chapter 2, the sources of the elements of the Freitian Variation were derived from the international community and treaties. The state's intention to provide a more cohesive legal framework conducive to the development of Taiwan's economy, international trade and maritime shipping industries can be found in the 1999 Maritime Act.

Notably, the amendments of the Maritime and Shipping Acts in the 1990s reveal that the state strove to align its relevant legal framework with international standards. In terms of the Steinbergist Classification, the ROC's construction of the ocean integrated elements of the Grotian Variation in its use of important international treaties with regard to seaborne trade and transport, as shown in Chapter 2. The critical momentum behind this adaptation and adoption was the state's intention to promote further economic growth by setting up the Asia-Pacific Operation Centre on Taiwan. This required adhering to the rules of the international community for utilising the marine surface to transport goods, which, in turn, required unrestricted navigation of the seas subject to weak forms of governance by the community of states and international organisations. Above all, by the 2000s, the ROC's social construction of the ocean regarding its maritime affairs fell within the parameters of the Mediterranean Sea Model and incorporated elements from both the Freitian and Grotian Variations as to the maritime activities of the society of modern Taiwan.

As to the extraction of resources from the ocean, especially fishery, the ROC also stipulated the Fishery Act in 1929. The Fishery Act and the 1929 Maritime Act together were the first step of altering the state's social construction of the ocean, although the construction generally remained in the Indian Ocean Model before the 1950s. When the state considered Taiwan's fishing industries (especially the deep-sea fishery) as a driver of further economic development, the Fishery Act was amended. As shown in Chapter 2, the 1971 Fishery Act aimed to regulate and manage the fishing industries, not to govern the seas' fishing grounds. Not until the 1991 Fishery Act, the 1998 Law on the Territorial Sea and the Contiguous Zone of the Republic of China and the 1998 Law on the Exclusive Economic Zone and the Continental Shelf of the Republic of China were promulgated did Taiwan consider the seas an arena that could be possessed by the state. The promulgations of the latter two marine-territorial laws signal that the state had absorbed elements of the Seldenian Variation and the Micronesia Model into its construction of the ocean.

The 1991 Fishery Act and the two marine-territorial laws were a response of the ROC to the declaration of Exclusive Economic Zones (EEZs) by its neighbouring states. The 1982 United Nations Convention on the Law of the Sea (UNCLOS) came into effect in 1994, under which the construction of the ocean according to the Grotian Variation was consolidated. As to the EEZs and contiguous zones, UNCLOS institutionalised the construction of the ocean consistent with the Seldenian Variation. With regard to the territorial sea, UNCLOS materialised a social construction of the ocean that was in alignment with the Micronesia Model (Steinberg 2001).

Having lost recognition as the sovereign representative of the Chinese nation, the ROC was prohibited from participating in UNCLOS. In order to protect Taiwan's legal and economic interests, the 1991 Fishery Act and the two 1998 marine-territorial laws were established. They were heavily influenced by UNCLOS and, hence, it can be argued that elements of the Grotian and Seldenian Variations, as well as the Micronesia Model, were incorporated into the state's social construction of the ocean by the end of the 1990s.

Thus far, I have summarised the trajectory of the changes in the ROC's construction of the ocean after 1949. In this trajectory, one of the main approaches that the state adopted to develop and sustain Taiwan's economic growth was the regulation of the use of the seas. The ROC's maritime and marine laws and the marine policy white papers were the manifestations of its representation of the ocean. Furthermore, the state controlled by the DPP also strove to foster what the DPP called an 'ocean-oriented culture' in the society of modern Taiwan, the cultivation of which also required further incorporation of the elements of the Micronesia Model into the state's construction of the ocean. Whether or not the state's efforts to develop a marine culture were successful remains to be seen. Likewise, together with the state's endeavours to seek economic development utilising the ocean, a main thesis of Steinberg's (2001) territorial political economy suggests that using (or intending to use) the ocean can influence the social construction of the ocean. Chapter 2 not only analyses the ROC's social construction of the ocean but also demonstrates the utility and applicability of the theory of territorial political economy to the case of the society of Taiwan between 1949 and 2016.

The Social Constructions of the Ocean by the Ideological Sources of Power in the Society of Modern Taiwan

Chapter 2 and the previous section suggest that the ROC led by the DPP attempted to nurture an ocean-oriented culture in Taiwan society. How the ideological powerholders in society responded to that pursuit is, however, an empirical question. Societal ideological powerholders could be independent from (though not necessarily against) the state's cultural initiatives. Chapter 3 delves into how modern Taiwan's ideological sources constructed the ocean through evaluating and interpreting the written works of some important societal intellectuals against the backdrop of the Steinbergist Classification.

For Mann (2012a), ideology is synonymous with culture, which comprises concepts and theories, norms and values and religious faiths and aesthetic judgements. Institutions and discourses centred on ideology are the ideological sources of power. People or organisations controlling institutions or creating and manipulating discourses are the ideological powerholders. Thus, Chapter 3 examines the literature produced by influential natural scientists, academics in the fields of economics and management, legal and political scholars and intellectuals in the arts and humanities in contemporary Taiwan.

For practical reasons, I cannot examine publications in every academic discipline. In the area of natural science studies of the ocean, I only interpret constructions of the ocean based on popular science writings and policy advisory papers. Chapter 3 suggests that marine engineers who urged the state and their fellow citizens to pay attention to the risk of terrestrial erosion by the ocean in coastal areas represented and constructed the seas as a source of danger. Their social construction of the ocean can be classified as the Indian Ocean Model.

In other natural scientists' written works that are discussed in this monograph, the same tendency can be found; that is, how one uses, intends to use, or urges others to use the ocean affects how the ocean is represented and constructed. Chapter 3 also suggests that marine biologists who advocated the protection of Taiwan's marine ecosystem and the preservation of the marine biodiversity of the surrounding seas constructed the ocean as the Micronesia Model. The same type of construction of the ocean can also be seen in the policy advisory papers by marine physicists who emphasised the need to further explore the deep sea and to develop the technology to conduct such explorations.

For these natural scientists, the ocean was represented as a resource reservoir: A geographical phenomenon that is plentiful and abundant rather than empty and void, full of both material (for example, fishes) and immaterial (for example, knowledge about deep-sea geophysical conditions) resources. Their construction of the ocean, thus, can be classified as the Micronesia Model, manifested in the following proposition: How natural scientists aimed, or urged others, to use the ocean influenced how they represented and constructed the ocean.

In the case of the literature on maritime and ocean affairs produced by economics and management scholars in Taiwan before 2016, diversified concerns can be found. These concerns are closely related to their various constructions of the ocean. Specifically, as shown in Chapter 3, managerial experts primarily focused on the increase and maintenance of the maritime shipping and commercial port service industries. Modern Taiwan's economics scholars concentrated on enhancing and adjusting Taiwan's fishing industries through the state's policy guidelines based on the relevant international regime rules. Moreover, economics papers regarding the uses and regulations of oceanic tourism and marine resources tended to highlight the abundant resources provided by the seas and the need for preservation through territorialisation and governance.

It is clear that the diversified concerns of different disciplines reflected scholarly advocacy for using and regulating the ocean, which, in turn, influenced their constructions of the ocean. This interpretation of the academic literature is consistent with the first main thesis in the theory of territorial political economy addressed earlier, which is also consistent with the views of legal and political scholars of modern Taiwan before 2016.

The third chapter of this monograph also examines and interprets influential legal and political scholars' written work. Legal scholars' papers and monographs related to maritime issues were not readily available in modern Taiwan. Although textbooks on maritime law for university students are an exception to the above statement, they are mainly introductory and educational in nature. However, marine affairs could still be considered an important issue amongst modern Taiwan's legal scholars. Public international law and UNCLOS provide the explicit framework for these studies in the literature. Legal scholars' publications on ocean-related topics typically investigated and evaluated issues related to sovereignty and legal disputes over the demarcation of ocean zones.

The works of modern Taiwan's political scholars writing before 2016 can be divided into three sectors. The first deals with marine affairs, examining the political actions and processes behind the sovereignty issues and legal disputes over marine-territorial delimitations. Although these scholars studied the politics of the disputes in the context of Sino-American or Sino-Japanese relations, UNCLOS and the territorialisation of marine areas were still taken into account in their analyses.

The second sector that copes with maritime affairs generally analyses sea power, including issues such as naval strength, geopolitical strategies, military tactics and naval-merchant co-operation. The third sector considers policy advocacy. Scholars of this area generally provided policy initiatives or prescriptions for what they considered problems in the field of maritime and ocean affairs. Nonetheless, the scholars focusing on these issues also followed the rationale of UNCLOS.

Above all, legal and political scholars who followed UNCLOS to develop their analyses generally constructed the ocean as a possessable resource reservoir. Their written works generally advocate the utilisation of political ocean resources, and some even support the protection and preservation of these resources by the state. This approach constructed the ocean consistent with the Micronesia Model. In the same vein, political scholars who analysed and urged the state to develop sea power constructed the ocean in the manner of the original Mediterranean Sea Model. Legal and political scholarly writings in the society of modern Taiwan illustrate the aforementioned first thesis in the theory of territorial political economy (Steinberg 2001).

My discussion of the literature by modern Taiwan's maritime historians and literary writers also exemplifies the first thesis mentioned above. I have dedicated much space in Chapter 3 to a survey of their works, and it is not necessary to repeat here the details of how they represent the ocean. I will, however, summarise the main points that are useful for this concluding chapter. Modern Taiwan's maritime historians generally constructed the seas in alignment with the Indian Ocean Model. Modern Taiwan's marine literary writers, by and large, constructed the ocean in the same pattern, with the exception of the works by a Tao writer, who constructed the thalassic according to the Micronesia Model.

Historians studied how maritime trade and migratory activities were undertaken mainly during and after the Ming dynasty (1368–1644) on the Chinese mainland, Taiwan and overseas. Their accounts do not analyse what occurred on the seas or any events related to the ocean itself, but rather the trade and migratory lives on the lands across the ocean. The view of the ocean *per se* of the people analysed by the historians was an empty void and spatial surface that had to be traversed to conduct trade and to pursue a better standard of living through migration. The historians' accounts of how people conducted maritime travel and surmounted marine barriers for trade and migration were effectively their own representations of the seas. Thus, modern Taiwan's maritime historians' constructions of the ocean were affected by how they accounted for others' uses of the ocean. Their constructions of the ocean can be classified into the Indian Ocean Model.

In the case of marine literary writings, there were few famous and well-received writers. One of the most well-known is Syaman Rapongan, a Tao marine literary writer. For Syaman Rapongan and the Tao people living on Lanyu, the marine-human relationship is harmonious. Tao's culture is marine-laden, and the lives of its people are integrated with the ocean. His description captures the viewpoint of traditional Tao people, in which the ocean is plentiful and lively. The ocean also provides daily necessities for a traditional Tao lifestyle and was constructed by Syaman Rapongan in the fashion of the Micronesia Model. Marine literary works by ethnic Han Chinese writers on modern Taiwan depict and construct the ocean differently. For example, in LIAO Hong-chi's works, the ocean is a remote spatial surface where marginalised people could fish for a living and could hide from the complex, exhausting social constraints and difficulties of life on land. In their novels, the ocean is presented as pristine and simple, in contrast to the corrupt and difficult land. Similar to the Western romantic oceanic literature in the era of industrial capitalism roughly between the 1760s and 1960s, the ocean in Taiwan's Han Chinese marine literature becomes 'the antithesis of land-space' (Steinberg 2001: 113). This construction of the ocean is consistent with the Indian Ocean Model.

To recap, modern Taiwan's historians, who aimed at representing how people, particularly in the areas of southern China, used the seas to undertake trade and migration effectively, constructed the ocean as did the people that the historians described and discussed in their historical accounts. That is, the historians' works construct the ocean corresponding to the Indian Ocean Model, which is also applicable to the literary works by modern Taiwan's ethnic Han Chinese writers. The construction of the ocean as the antithesis of land and society is aligned with the Indian Ocean Model. The writers explicitly or implicitly show their normative values of human-marine relations and urge others to interact with the ocean in ways that are consistent with their values. This proposition also reflects Syaman Rapongan's novels even though they construct the ocean consistent with the Micronesia Model, explicitly praising traditional Tao values and harmonious-patterned interactions with the ocean and implicitly urging the Tao people to respect and restore their traditional lifestyle.

Hence, the written works of the arts and humanities intellectuals in the society of modern Taiwan surveyed in the third chapter demonstrate the first main thesis of the theory of territorial political economy presented in the previous sub-section. Chapter 3 interprets the literature produced by the ideological sources of power in Taiwan through the lens of the Steinbergist Classification, and, together with Chapter 2, illustrates and exemplifies the thesis that how the ocean is used has a direct influence on how the ocean is socially constructed. In other words, the use, regulation and representation of the ocean are mutually interrelated, and their interactions shape the construction of the ocean.

One additional point worth mentioning before concluding this sub-section is that, as I suggest in Chapter 2, the ROC, especially under the DPP's reign between 2000 and 2008, attempted to cultivate an ocean-oriented culture in Taiwan. What this term means has not been clearly defined. However, according to the two marine policy white papers published in 2001 and 2006 (Council of Marine Affairs Advancement 2006; Research, Development and Evaluation Commission 2001), it can be inferred that such a marine culture would be close to the Micronesia Model. Nonetheless, as Chapter 3 has demonstrated, societal ideological/cultural powerholders had their own agenda. Indeed, it has become a cliché to claim that it remains to be seen whether or not an ocean-oriented culture can be successfully fostered in Taiwan. Nevertheless, what is important, I suggest, is that how the state interacts with the societal ideological powerholders to promote this cultural initiative is the essence of the politics of marine policy development and implementation.

The Social Construction of the Ocean by the Military Sources of Power in the Society of Modern Taiwan

Chapter 4 examines how the ROC Armed Forces constructed the ocean between 1949 and 2016. This chapter is divided into three time phases: From 1949 to 1971, from 1971 to 1991 and from 1991 to 2016, which form the cases for the chapter's concluding comparative discussion. Chapter 2 and 4 demonstrate the aforementioned first main thesis, and Chapter 4 illustrates the second thesis with the concluding comparative discussion. The comparative discussion supports the second thesis, which indicates that the condition of a geographical phenomenon can influence its related organisation of society and, in turn, the social construction of that geographical phenomenon.

The narrative analysis presented in Chapter 4 explicates how the ROC's military forces used the ocean and its relevant historical experiences. It interprets the use and experiences in light of the Steinbergist Classification, which consists of the academic contribution in its own right. Moreover, the contribution of such an interpretation has the merit of enabling the comparative discussion mentioned above. Thus, as in Chapter 5 (which I summarise later), the narrative analysis in Chapter 4 serves two functions: Comprehending the construction and demonstrating the theory of territorial political economy.

The ROC's military forces used the ocean according to the state's grand strategies, which provided the guidelines for its Armed Forces' military strategies. As Biddle (2007) identifies, a grand strategy represents the state's ultimate security goals, including the ideological, military and economic approaches to achieve the goals. Moreover, a military strategy prescribes how a state's military should fulfil the role assigned by the state's grand strategy and achieve the state's ultimate security goals. It typically develops the actions of the largest military units and formations. (Biddle 2007). Thus, interpreting the ROC's grand strategy, the strategies of its military force and their historical contexts through the lens of the Steinbergist Classification allows for an understanding of the social construction of the ocean by the ROC Armed Forces; it also establishes the basis for demonstrating the aforementioned second main thesis in Steinberg's (2001) theory of territorial political economy.

From 1949 to 1971, JIANG Jieshi (also known as CHIANG Kai-shek; hereafter JJS) still viewed the ideological differences between the KMT and CCP as a zero-sum, life-or-death struggle. He strove to implement the ideals of the Three Principles of the People (TPP). After Jiang led the withdrawal to Taiwan, the ROC retained around 600 thousand troops; maintaining military power consumed approximately 70 per cent of the annual national budget. It was imperative for JJS and the ROC to promote economic recovery, growth and industrialisation. Meanwhile, JJS consolidated his domestic authority within the KMT and the state organisations. Geopolitically, he managed to secure the military protection of the United States (US). JJS also deployed approximately 150 thousand soldiers on Jinmen and Mazu islands, poised to make a counterattack against the PLA on the Chinese mainland.

According to JJS and the White Group's Plan of Glory, by 1958, the ROC's military power in Taiwan would have been ready for the counterattack. The ROC Navy (ROCN) would have been in charge of securing the command of the Taiwan Strait and transporting military personnel and material. Although the Plan of Glory was never implemented, the ROCN was assigned the responsibility of consolidating its military domination of the sea and transporting special operation forces for amphibious assaults on coastal areas of the Chinese mainland. Notably, between 1949 and 1965, with US assistance, the military balance in the Taiwan Strait between the ROC's military forces and the PLA still leaned towards the former. Under such circumstances, and with the grand and military strategies focused on retaking the mainland by force, the ROC's military socially constructed the ocean in the manner of the original Mediterranean Sea Model.

By 1965, JJS realised that it was improbable, if not impossible, to restore the ROC's rule over the Chinese mainland. The grand strategy and its corresponding military strategy changed and focused instead on the military defence of Taiwan Island. This position continued into the period between 1971 and 1991. The ROC lost its recognition by the United Nations in 1971 and JJS passed away in 1974. The geopolitical relationship between the ROC and the US was transformed from one of interdependence into one of dependence by Taiwan on the US. Domestically, the ROC led by JIANG Jingguo (also known as CHIANG Ching-kuo; hereafter JJG) achieved a high rate of economic growth and industrialisation during this period. In the later years of his administration, JJG also began to liberalise the society of Taiwan and planted the seeds of democratisation of the ROC. The process of liberalisation and embryonic attempts at democratisation not only alleviated the political demands from the US but were also intended to curb the increasing power of the TIM and to counterbalance the CPP's political (and diplomatic) pressure. Although JJG and the KMT continued to uphold the TPP against Chinese Communism, launching the first military strike against the PLA and the CCP to restore ROC governance on the mainland was out of the question.

The military forces would not be used to project the ROC's power onto the mainland nor be deployed for commanding the Taiwan Strait. Its military strategy instead was to defend Taiwan against possible saturation attacks by the PLA. Therefore, the ROC Air Force (ROCAF) was assigned to secure air supremacy, and the ROCA was tasked with repelling the PLA's landing attempts. The ROCN was ordered to intercept the PLA's follow-on amphibious and logistic forces. The ROCN also had the mission of ensuring the safety of Taiwan's eastward sea lane of communication with the US when the military conflict commenced. Thus, when the ROC's grand strategy no longer had an offensive element and the military balance in the Taiwan Strait leaned towards the PLA, resulting in the ROCN's inability to command the strait, the ROC's military sources of power constructed the ocean in a way that was aligned with the Indian Ocean Model. The seas were no longer considered a military arena. Rather, it became a source of an amphibious threat and danger.

The ROC's grand strategy changed again in the early 1990s. JJG died in 1988, and LI Denghui (also known as LEE Teng-hui; hereafter LDH) succeeded the presidency, consolidating his political power in both the KMT and the ROC. Domestically, LDH promoted the democratisation of the state and the co-operation of the members of the TIM. He became the first popularly elected president of the ROC and announced his Two-state Theory in 1999, which became the primary ideological component of the state's grand strategy. The state was linked with the TIM through LDH's endeavours.

Democratisation continued into the late 1990s, culminating in CHEN Shuibian (also known as CHEN Shui-bian; hereafter CSB) winning the presidency in 2000. As a DPP member, CSB maintained the ROC's linkage with the society's TIM. Effectively, since the late 1990s, the cross-Strait relationship was deteriorating. Both LDH and CSB believed it was essential to minimise Taiwan's economic ties with the Chinese mainland if ROC co-operation with the TIM was to be sustained and strengthened. They thus promoted Taiwan's economic globalisation, excluding the Chinese mainland market (Leng 2016).

The military strategy of the ROC Armed Forces during the terms of these two presidents also differed from that under JJG's leadership. The military strategy actually began to change gradually, starting in the early 1990s. The ROCN and ROCAF were ordered to deal with the PLA's blockade of Taiwan Island, a shift in their previous responsibility to intercept the PLA's amphibious offence. The military forces of the ROC underwent modernisation in the 1980s and received new weaponry during the presidencies of LDH and CSB.

The military believed the PLA was much less likely to launch a direct, confrontational assault through an amphibious landing. Furthermore, the alteration of military strategy corresponded with the ROC's strengthened connection with the TIM and the deteriorating cross-Strait relationship. Confident with their modernised weapons and related tactics, the ROC transformed its military strategy from a terrestrial approach aimed at expelling the invasion on the shore of Taiwan Island into a maritime-centred approach deploying powerful naval flagships in critical marine areas as pivotal units to coordinate the ROCN and ROCAF forces and to defeat the enemy in the Taiwan Strait or to retaliate against targets on the Chinese mainland. The ocean became a military arena for the military sources of power wielded by the state, socially constructed in accordance with the original Mediterranean Sea Model.

By 2008, the KMT candidate MA Yingjiu (also known as MA Ying-jeou; hereafter MYJ) won the presidential election and led the ROC between 2008 and 2016. During his period of governance, he attempted to separate the state from the TIM. The ideological aspect of the state's grand strategy no longer emphasised hostilities between Taiwan and the Chinese mainland. Instead, it encouraged people of Taiwan to concentrate their energy on economic development and attempted to ameliorate the cross-Strait relationship. The economic aspect of the grand strategy coherently endeavoured to advance Taiwan's development through the Chinese mainland market (Leng 2016).

At the same time, democratic governance continued and was consolidated from the time of the inauguration of the MYJ presidency in May 2008. Additionally, the cross-Strait relationship became a political competition over the legitimate origin of state authority rather than an urgent geopolitical conflict. This was reflected in MYJ's Diplomatic Truce, which was based on his acceptance of the One-China Principle. Acceptance of the One-China Principle was the bottom-line requirement by the CCP for political parties and people across the Taiwan Strait to undertake interactions and exchanges with the mainland.

Meanwhile, by 2008, the ROC's Armed Forces were well aware of the military modernisation that had been undertaken by the PLA. In order to better protect the ROC, the state's military forces redesigned their strategy. As mentioned in Chapter 4, this modernisation followed the rationale of the porcupine strategy (Bush 2016; Murray 2008). The MYJ administration adopted a hedging approach to deal with the ROC-PRC relationship and that between the ROC and the US. The ROC created a geopolitical environment that allowed its military forces to readjust. The Armed Forces reverted to a military strategy that was similar to that developed during JJG's rule, but that placed more stress on all the military forces conducting joint defensive operations.

In this military strategy, air supremacy, command of the sea and ground force dominance were expected to be obtained jointly. Thanks to MYJ's improvement of the cross-Strait relationship and the geopolitical strategy of hedging, the ROC's military could concentrate once again on the main island of Taiwan, fortify its missile system and adopt a passively defensive posture. Under such a condition, the ROCN was no longer focused on defeating the enemy at sea but was rather integrated into the overall strategy of taking advantage of Taiwan's geography and the Taiwan Strait as a barricade to exploit the naval weakness of the PLA. The ocean, hence, once again represented a source of possible danger and threat and not an arena for pursuing decisive naval battles. The decisive military showdown would have to wait until the PLA reached the shores of Taiwan Island. The ocean was constructed by the ROC's Armed Forces during this period according to the Indian Ocean Model.

Chapter 4 interprets the history of the ROC's grand strategy and its Armed Forces' military strategy between 1949 and 2016. Chapter 4 is divided into three periods and provides two comparisons. Through these two comparisons, the aforementioned second thesis of the theory of territorial political economy is demonstrated. That is, the comparisons illustrate that nature influences the material organisation of society and, in turn, the social construction of the ocean.

Two structural factors stand out in Chapter 4: The state's grand strategy and the balance of military power between the ROC Armed Forces and the PLA in the Taiwan Strait. The Taiwan Strait influenced modern Taiwan's grand strategy and military strategy, which in turn affected the military forces' construction of the ocean. The Taiwan Strait is wider than the English Channel and notoriously difficult to cross, especially during times of war. It is known that between 1949 and 2016, the PLA did not have sufficient transport capacity to cross the strait and launch an amphibious assault.

The first comparison consists of three phases (as cases): From 1949 to 1971, from 1971 to 1991 and from 1991 to 2008. In the first phase, the ROC's grand strategy involved an invasion of the Chinese mainland and its stronger navy, vis-à-vis PLAN, managed to command the seas (before 1965). In the third case, the ROCN was confident in its ability to destroy the PLAN at sea, and the state's grand strategy involved a coercive, aggressive posture in deterring the PLA. In contrast, in the second case, the ROC's grand strategy abandoned the goal of restoring its government to power on the Chinese mainland. Instead, the Armed Forces' military strategy focused on defeating the PLA on the shores of Taiwan. Thus, when the ROC adopted an aggressive military stance and possessed a comparatively stronger naval *qua* military power, the construction of the ocean by the ROC's military forces became that of the original Mediterranean Sea Model. When the ROC undertook passive, defensive grand and military strategies and the military balance did not favour the ROC Armed Forces, the military constructed the ocean consistent with that of the Indian Ocean Model.

The second comparison also consists of three phases (as cases): From 1991 to 2000, from 2000 to 2008 and from 2008 to 2016. In the first and second cases, the ROC adopted an aggressive, coercive grand strategy against the

PRC and the PLA, even though its military strategies still prepared for second strikes and unprovoked invasions. Meanwhile, the ROCN retained its confidence in defeating the PLAN in the Taiwan Strait and possessed modernised weapons to facilitate its maritime ambitions. In contrast, from 2008 to 2016, the ROC's grand strategy aimed at fostering a friendly relationship with the Chinese mainland through MYJ's acceptance of the One-China Principle, Diplomatic Truce and geopolitical hedging policy. The corresponding military strategy refocused on the military defence of the island of Taiwan. Thus, similar comparative results can be reached. When the ROC's grand strategies were inclined to be offensive and its Armed Forces enjoyed an upper hand on the sea, the military's construction of the ocean fell in line with that in the original Mediterranean Sea Model. When the ROC's military no longer had a naval advantage over the PLA and its strategy switched to a defensive one oriented towards building an amicable relationship with the CCP, the construction of the ocean can be classified as that of the Indian Ocean Model.

The Social Constructions of the Ocean by the Economic Sources of Power in the Society of Modern Taiwan

The second main thesis of the theory of territorial political economy is presented in Chapter 4. The third thesis is demonstrated in Chapter 5, in which the development of the two most prominent marine-related sectors of the economic powerholders in modern Taiwan, the maritime shipping industries and the fishing industries, are discussed. The discussion of their development is arranged in a thematic way, which shows how they socially constructed the ocean.

In the case of Taiwan's maritime shipping industries, they socially constructed the ocean in line with the Indian Ocean Model; they did not consider it to be in their interests to deem the ocean possessable, which might have harmed their freedom of navigation and hindered their use of the ocean surface for cargo transport. Judging how they used the ocean, they did not consider the seas as resource reservoirs. The resource they needed was the marine spatial surface conducive for maritime transport. The Micronesia Model is therefore excluded.

The reasons for classifying the social construction of the ocean by Taiwan's maritime shipping industries as the Indian Ocean Model can be found in their actual use of and preparation for the use of the ocean, which sought to maximise profits, minimise costs and maintain the competitiveness of their operations. Certain key aspects or themes were crucial for the development of a competitive advantage, according to the autobiographies of some of the chief executive officers of certain large preeminent maritime transport companies, including Evergreen Marine and Yang Ming Marine Transport.

The first theme involves ships. Although this may appear to be self-evident, possessing sufficient ships of good quality was the first necessary condition for the competitiveness of a maritime transport enterprise.

The functions of ships were critical as well. Purchasing or building ships for the required functions and quality could increase the efficiency and profit of the maritime transport operations. For example, Evergreen Marine deployed ships designed for expansion during the oil crisis; this expandable attribute allowed them to increase carrying capacity when international trade started to grow. Flexibility meant efficiency and profit. Containerisation was another important measure to increase transport efficiency and a necessary step for a maritime transport company to compete globally. Experience shows that the maritime transport companies that lacked ships with the required quality and functions had to rely on loans or state subsidies. Financial problems would haunt those companies unless they were able to obtain the necessary amount of money to acquire the ships they needed, as in the case of Evergreen Marine's dependence on Marubeni's financial assistance.

The second theme is cargo. Deploying a sufficient number of suitable ships was one of the means to obtain more cargo and earn greater profits. Using specialised ships for some cargo that has special requirements during transport could increase efficiency and lower costs. Moreover, as mentioned above and in Chapter 5, employing container carriers was also a way to improve efficiency and reduce the cost of cargo transport. Higher efficiency and lower costs meant cheaper freightage and more opportunities to obtain more cargoes. Another way to acquire cargoes was by providing deep-sea liner shipping services sailing the routes where there was high demand for more and better maritime transport services. Certainly, to know where the demand was high required market surveys by the maritime transport companies. Furthermore, to acquire additional cargo, maritime transport companies, such as Evergreen Marine, even assisted cargo owners in expanding their market share in other countries to increase the need for shipping services. Additionally, container carriers were getting larger, sailing longer and entering fewer harbours. Deep-sea liner shipping corporations began to coordinate or cooperate with near-sea transport companies. Maritime transport enterprises also established international shipping alliances and joint operations. These arrangements allowed the maritime transport enterprises to allocate their shipping capacity and cargoes more economically, thus earning more profits.

The third theme is terrestrial logistic support for maritime transport. One of the most important types of logistics support provided on land, as identified by some chief executive officers of powerful maritime transport companies, was human resource recruitment, training and management. This type of logistic support was critical for the survival and development of the companies. Each of the influential maritime transport corporations discussed in Chapter 5 made great efforts to cultivate and improve their human resources. Important logistics support could also include building and maintaining terrestrial shipping networks for containerised transport. Examples include liner shipping agency networks, near-sea maritime transport networks for cargo transitions and terrestrial container and cargo transport networks.

For example, Uniglory Marine Corporation assisted Evergreen Marine and handled intraregional cargo transition. Everglory Transport Corporation handled inland container haulage. These subsidiary companies enabled Evergreen Marine to control the quality of its entire chain of shipping services.

In the above summary of the first part of Chapter 5, I have suggested that the maritime shipping industries of modern Taiwan did not construct the ocean according to the Micronesia Model. It should now be clear from the above summary that the industries' construction of the seas was more aligned with the Indian Ocean Model. That was because their operations and actual use of the ocean concentrated on maximising profits, minimising costs and maintaining the competitiveness of their maritime shipping services. The ocean *per se* was relegated to a secondary role, which provided a spatial surface for the industries to cross and deliver goods. Furthermore, as also elaborated in Chapter 5 of this monograph, modern Taiwan's maritime transport enterprises might have been better off without the state's compulsory military transport services and shipping of strategic resources, which usually pushed the service providers into financial deficits. Such practices caused operational and even solvency difficulties. For Taiwan's maritime transport corporations, state interventions were something to be avoided. This was hardly a construction of the Mediterranean Sea Model or the Freitian Variation.

The second part of Chapter 5 interprets the uses of the ocean and explicates the constructions of the ocean by the four major branches of the fishing industries of modern Taiwan. That chapter discusses the fisheries' constructions of the seas and demonstrates the third thesis of the theory of territorial political economy. Steinberg (2001) posits that a person's or group's construction of the ocean can influence the relevant material organisation of the ocean and, in turn, can affect the geographical condition of the ocean. The illustration of the thesis can be seen in my explication of the history of the fishing industries in Chapter 5 and the summary below.

The fishing industries can be divided into four branches: Aquaculture, near-shore fishery, near-sea fishery and deep-sea fishery (Lee 2007). In modern Taiwan, aquaculture is made up of inland and offshore aquaculture. By examining their practices and what was important for people undertaking these types of fisheries, it is possible to interpret their constructions of the ocean. Throughout the history of the development of modern Taiwan's aquaculture, both inland and offshore, access to technology to breed and raise suitable species to generate revenue was crucial, attracting much attention from the fish farmers and concerned state officials.

Inland aquaculture, however, was aggravating land subsidence, especially in the coastal areas of Taiwan Island. For inland aquaculture in the coastal areas, land subsidence could result in seawater intrusion and the erosion of usable land. The ocean was thus a paradox for inland aquaculture. On the one hand, the seas posed a serious threat to the continuity of coastal inland aquaculture. On the other hand, seawater extracted from the ocean could be used for aquaculture to minimise reliance on underground freshwater. Thus,

I suggest that modern Taiwan's inland aquaculture constructed the ocean as does the Indian Ocean Model.

In the case of offshore aquaculture, its social construction of the ocean followed the Seldenian Variation. Offshore aquaculture was generally undertaken by relatively small fish farmers instead of large enterprises, which were effectively discouraged by the state. The ROC territorialised the surrounding sea areas of Taiwan in the 1990s, and the state deemed it imperative to govern the fishing grounds in its EEZ. The state granted fishing rights to fishers and fish farmers according to the Fishery Act.

Small-scale fish farmers undertaking offshore aquaculture utilised fixed cage nets, thus occupying a certain area of the sea to raise fish. The state's Seldenian construction of the ocean embedded in the Fishery Act, as shown in Chapter 2, had an impact on the marine fish farming industry's construction of the ocean. In other words, the actual use of the ocean by offshore aquaculture practitioners in the society of modern Taiwan, combined with the ROC's influence through the regulation of the Fishery Act, effectively constructed the ocean in the style of the Seldenian Variation.

Modern Taiwan's near-shore fishery constructed the ocean in the manner of the Indian Ocean Model. In the case of the near-shore fishery, the ROC's own changing construction of the ocean exercised little, if any, influence on that of Taiwan's near-shore fishery. How the ocean was constructed previously can be traced back to the pre-modern era of Taiwan's agricultural economy before Japan's colonisation in 1895. Nevertheless, this fishery's condition and construction of the ocean in the modern era after 1949 was related to Taiwan's overall economic situation.

The near-shore fishery's practices of ocean utilisation were a dominant factor influencing its construction of the ocean. Although Taiwan's near-shore fishing industry remained primitive, requiring low capital investment and technology, it was the primitivity that could absorb the unemployed and provide a source of income for fishers and the areas surrounding fishing villages after the Second World War and in the early 1950s. Moreover, although the ROC promoted near-sea and deep-sea fisheries after the mid-1950s, the near-shore fishing industry functioned as an economic safety valve functioned, and its production began to grow again in the 1970s when an increasing number of states claimed their EEZs and the oil crisis hit the global economy.

The near-shore fishery industry did not view fishing grounds or the seas as private property. The sole focus was on earning a living. The ocean was considered a spatial surface, not as anyone's property. On the surface, catching fish, which seemed to be an unlimited resource, raised no environmental or resource conservation concerns. Fishing was merely a way for the marginalised of society to earn a living, representing a construction of the ocean according to the Indian Ocean Model.

Modern Taiwan's near-sea fishery industry initially followed the same pattern as that of the near-shore fishery, but its construction changed mainly in the 1990s, absorbing elements of the Seldenian Variation. Both the near-sea and near-shore fishery industries caused the problem of overfishing. This environmental threat was a common by-product of the utilisation of the ocean following the Indian Ocean Model. Overfishing also occurred in the case of modern Taiwan's deep-sea fishing industry (which I discuss later). In this model, the ocean belongs to no one and simply is reduced to a surface and a void. In this model, the externalities of economic activities caused by the marine users are indifferently attributed to the marine space. All the attention and energy are focused on the profitability of the economic activities.

The Japanese colonial government first facilitated the development of Taiwan's near-sea fishery industry. By the mid-1950s, the ROC also encouraged near-sea and deep-sea fisheries. As I have suggested in Chapter 5, the state's promotion of the fishing industries and people's involvement in economic activities related to fishing aimed to increase production and maximise profits by utilising the marine surface to fish. The locations for catching fish or the sea *per se* were of no concern. This point can be illustrated by the ROC's promotion and fishers' use of new equipment and methods. Marine areas with fishing potential were explored and exploited. The new equipment, methods and fishing grounds were utilised with little concern for the deterioration of the marine and fishery environment.

The Indian Ocean construction of the ocean by the near-sea fishery remained largely intact until the 1990s, with its production reaching its peak in 1980 (Lee 2007: 69). The construction assimilated elements of the Seldenian Variation resulting from the EEZ of the ROC overlapping with those of Taiwan's neighbours. Conflicts of interest in these overlapping EEZs led to the harassment of Taiwan's near-sea fishermen by other countries' police and military forces and fishing vessels. The conventional fishing grounds of Taiwan's fishermen were also facing environmental threats. Although the demand for protection and regulation by the state did not completely shift the near-sea fishing industry's construction of the seas away from the Indian Ocean Model, the construction did incorporate some elements of the Mediterranean-Seldenian Variation.

Modern Taiwan's deep-sea fishery also constructed the ocean in the pattern of the Indian Ocean Model at first, similarly to the near-shore and near-sea fisheries mentioned above. However, its situation was comparatively more complicated, resulting from its Indian Ocean-styled construction of the seas. Unlike the near-shore fishing industry, the deep-sea fishery required heavy capital and technological investment by private companies, even though the ROC encouraged its development after the mid-1950s.

Therefore, it was logical that the deep-sea fishery industry concentrated on profit-making. Any approaches that could bring in more return on investment would be welcome by the deep-sea fishing corporations. As Chapter 5 describes, corporations accepted state loans and focused its initiatives on building larger purse seiners and longliners. Deep-sea trawling ships were also deployed. By the late 1960s, the deep-sea squid jigging and ultra-low temperature tuna lining industries prospered. Furthermore, new fishing grounds and species were explored and exploited, for example, the fishing grounds in the Atlantic Ocean near South America.

Because the deep-sea fishery needed capital and technological investments, the industry was more similar to other terrestrial industrialised economic activities. I suggested in Chapter 5 that the deep-sea fishery was analogous to hunting in the field when its fishing vessels attempted to locate suitable fishing grounds and to mining in that it extracted fish from underneath the marine surface. The deep-sea fishery's initial construction of the ocean can be shown in its preparation for operations on the seas.

First, well-designed fishing vessels of good quality were needed. Second, suitable fishing methods and techniques that corresponded to the targeted fish and deployed vessels were required. Third, appropriate overseas bases for the deep-sea fishing operations were critical for logistic support, fish processing and efficiency and profit maximisation. Fourth, hiring a competent captain with the necessary leadership skills and fishing knowledge was essential for each deep-sea fishing vessel if its purpose was to successfully track, hunt and extract fish for profit. Fifth, appropriate initial processing of the harvested fish, adequate equipment and technology for transporting the catch and unimpeded access and well-connected networks for selling the harvests were additional requirements of the successful deep-sea fishing industry. Therefore, it is clear that the construction of the ocean by this fishing industry belonged to the Indian Ocean Model in approximately the late 1970s and early 1980s.

The Indian Ocean-fashioned construction of the ocean by Taiwan's deep-sea fishing industry was altered mainly because many states around the world established their EEZs. It is worth quoting here, as well as in Chapter 5, that EEZs 'were conceived of as spaces for implementing a Seldenian variant of stewardship' (Steinberg 2001: 171). For Taiwan's deep-sea fishing vessels to continue their operations in the jurisdictions of other states' EEZs, they typically had to pay for legal access or purchase back what they had harvested in the exclusive economic marine areas. The deep-sea fishing industry's payment of fees for legal fishing rights in EEZs, thus, facilitated the absorption of elements of the Seldenian Variation.

This transition took place in the late 1970s and early 1980s. Continuing to generate fishing externalities and actually overfishing ultimately caught the attention of the International Commission for the Conservation of Atlantic Tunas (ICCAT), which imposed stiff penalties in 2005. By this time, the ROC finally took serious action to respond to the externalities of modern Taiwan's deep-sea fishing activities. The government imposed strict regulations, adopted tough measures and spent significant financial resources to reform the deep-sea fishing industry of Taiwan. Taiwan's tuna fishing quota was restored by the ICCAT after the state and the fishing industry implemented reform measures. Thus, the deep-sea fishery, by and large, incorporated elements of the Grotian Variation into its pre-existing construction of the ocean.

Thus far, I have summarised how Taiwan's fishing industries constructed the ocean between 1949 and 2016. The related history and social experiences were interpreted through the Steinbergist Classification and introduced in Chapter 5. The interpretation also demonstrates the third thesis of the theory of territorial political economy. Importantly, the externalities generated by fishing activities of the Indian Ocean Model could be found in almost every sector of Taiwan's fisheries. The marginalised role of the ocean *per se* did not provide any incentives for the fishing industries, which did not view the ocean or the fishing grounds as their private property to utilise and conserve fishing resources. In different economic contexts and under different conditions, modern Taiwan's fisheries did not pay due attention to the environmental impacts of their fishing activities. Much energy and money were concentrated on increasing the profits of their fishing operations. Locating fishing grounds was similar to hunting and catching fish in the fishing areas was similar to mining.

As shown previously, the state enacted policies to encourage near-shore fisheries, and society created institutions to sustain the economy of near-shore fishing activities and villages. The state also promoted near-sea and deep-sea fisheries and stipulated laws and regulations to coordinate fishing activities. Societal economic powerholders could also establish firms and operate the related businesses to support these fishing industries. Furthermore, the capacity of modern Taiwan's fishing industries often led to overfishing. The deep-sea sector was punished by the ICCAT, whereas the problem of depleting fish stocks plagued the near-sea and near-shore fishing industries. Thus, the history of the development of modern Taiwan's fishing industries illustrates the influence of the social construction of the ocean on the material organisation of society and the geographical conditions of the ocean.

Bringing People Back In: The Micro-politics of Marine Policies

As I have mentioned in Chapter 1 and at the beginning of this chapter, the interpretation and introduction of the social constructions of the ocean by the four most important types of sources of social power in the society of modern Taiwan form a foundation of the structure of the political analysis of the development and implementation of Taiwan's maritime and ocean politics.

I suggest that when marine policies on Taiwan are investigated, it is necessary to explore the geographical conditions of the marine zones and the material organisations of society. For example, when one studies Taiwan's policy of offshore wind power, the influence of the natural attributes of the related sea areas and the relevant laws, institutions, organisations and technologies that are utilised to sustain offshore wind farms and power generation should be evaluated and considered as an important context. Likewise, it is also necessary to examine the uses, regulations and representations of the ocean by the actors involved. For example, in the case of offshore wind power, how the state regulates the ocean, how the offshore wind power enterprises represent the sea and how the fishing industries use the marine areas can be considered the social structural factors that affect how the political processes amongst these actors that underlay offshore wind power policy.

To make the above suggestions by no means, is to take sides in the structure-agency polemic. Instead, following Mann (2012b), I adopt an eclectic stance on the issue, emphasising the importance and usefulness of evaluating and comprehending the political intentions and interactions of actors against the structural contexts. As I show below, mainstream academic studies on marine policies in Taiwan and other international venues generally ignore the roles played by key actors, especially societal elites and political leaders.

Above all, as I have claimed in Chapter 1, studying the social constructions of the ocean by various critical power sources in Taiwan between 1949 and 2016 precipitates a foundation for future studies of the politics of marine policy. These future studies can provide detailed and balanced research results because they will allow investigations of the political processes of marine policy conditions and implementation. They will also take actors' intentions and interactions and the socio-structural factors into consideration.

Later, I follow Adrian Leftwich and Steve Hogg (2007; 2008), Mann (2012b) and Steinberg (2001) to establish a preliminary analytical framework for future research on the politics of marine policies in Taiwan. I next discuss my survey of the only academic journal of marine policy in Taiwan.

The Journal: *Marine Affairs and Policy Review*

This section attempts to show how marine policies are usually studied academically in modern Taiwan. I report my finding after surveying the only academic journal, written in Chinese, on Taiwan and dedicated to marine policy studies and advocacy: *Marine Affairs and Policy Review*. This journal is selected not only because it is the only available academic journal but also because it is published by the Institute of Marine Affairs and Policy, ROC, whose members include many important societal elites in the field of maritime and ocean affairs of Taiwan. Many of them are influential scholars, including Professors WANG Kuan-hsiung, SONG Yann-huei, CHIAU Wen-yan, LIU Fu-kuo and HU Nien-tsu. Other societal elites, including CHEN Zhulong, the chairman of Mingda Fishery Corporation and TAN Chih-lung, a retired rear admiral of the ROCN, are also key members of the institute. These scholars and elites contributed papers on marine policy to the journal. Thus, a survey of this academic journal can reveal how marine policies are studied and advocated in Taiwan's academia.

Nevertheless, the term 'policy' must be defined first. According to Mark Zacharias (2014), a policy is a plan or course of action that can affect, and even determine, subsequent decisions and actions. A policy is, effectively, a path to direct strategic and tactical orientations and objectives to govern complex and controversial problems. Similarly, the term 'marine policy' endorsed by the journal denotes a type of state policy that establishes the state's course of action regarding governing maritime and ocean affairs.

In this section, I will not dedicate space to introducing the principal argument of every publication in that journal. Nor will I quote and refer to any specific article, except that of Wen-yan Chiau (2017). However, the papers can be categorised into various themes. My brief introduction to the themes demonstrates that Taiwan's academic work on marine policy does not use an integrated perspective (Balgos, Cicin-Sain and VanderZwaag 2015; Zacharias 2014), which I briefly discuss later. Rather, almost all of the articles in the journal adopt a sector-based approach, each concentrating on an issue of maritime and ocean affairs of interest to the authors. With the exception of Chiau (2017), who advocates for the ROC strengthening its coordination of the related requirements and the implementation of marine policies, none of the authors discusses the utility of seeing marine policies as a whole, and all of them analyse their issues at hand in a legal-institutional style.

From the journal's first publication in 2011 to the latest available volume in 2019, the first theme expressed in the articles relates to the ROC's coast guard administration and its law enforcement on the seas. Each paper pays attention to different aspects of maritime law enforcement by the ROC coast guard in this theme. The notion of maritime security is mentioned occasionally, as is UNCLOS. However, what maritime security means is not clearly defined in the articles. The basis of their arguments is usually rooted in UNCLOS. Their arguments begin with this treaty and analyse how it has been implemented by the ROC to protect Taiwan's maritime and marine interests. The normative values of the authors are expressed through the authors' prescriptions for current situations. Additionally, a second discussion trend regarding the internal organisation of the coast guard administration can also be seen in these papers. It is worth noting that these discussions are all presented with legal discourses and analyses.

A second theme closely related to maritime law enforcement and security is maritime delimitation, especially in the areas of the East and South China Seas. Papers on this topic analyse confrontations and controversies of maritime delimitations, which have strong geopolitical implications, and the potential and present economic benefits. Although the geopolitics and the economy of the ocean are effectively embedded in the papers, political and economic concepts and theories play no major role in the articles. Once again, UNCLOS is the basis of the analyses of the current practices of the concerned laws and institutions. Institutional, organisational analyses are presented when domestic responses to the questions of maritime delimitations are considered. Historical explications of states' interactions centred on UNCLOS and the related international laws are offered (even though UNCLOS and international law were used as a pretext of the involved states' geopolitical actions) in the analyses of international disputes regarding maritime delimitations.

Third, the journal has six papers elaborating different authors' views on Taiwan's fisheries and fishing policies. A common pattern of argument that emerges in these papers is an introduction of the current situation of Taiwan's fishing industries, followed by a discussion of the lack of environmental protection and conservation by the state and the industries. They continue with an explication of the legal procedures and policies for conserving fishing resources available internationally and domestically. The concluding remarks of these papers generally refer to the sustainable development of fishing and other marine resources, urging the establishment and further improvements to existing legal procedures and regulations for sustainable development. Additionally, two related papers published in the journal discuss the functions of the policies of marine spatial planning and marine ecosystem management (if they can be properly implemented) and clarify how these policies relate to UNCLOS. However, their approach to the introduction of these policies is similar to the six papers mentioned above in that they are legal-institutional in nature, focusing on laws and regulations *per se*.

The fourth and fifth themes of the journal are tourism and recreational fishery and the underwater cultural heritage of modern Taiwan. Two papers deal with the former issue, and three address the latter. Those that cover tourism and recreational fishery describe its development and current challenges and offer legal and institutional prescriptions for the state to address the problems. For them, the problems are rooted in existing state policies, regulations and laws.[2] The papers on underwater cultural heritage introduce the relevant international laws and regimes. Their policy analyses fall within the discipline of law, and their policy proposals derive from existing legal systems. They seek to align the ROC's legal system with the related international laws and legal practices.

The sixth and seventh themes relate to the economic benefits of utilising the ocean: Maritime transport and shipbuilding. Six papers can be found dealing with these issues. Compared to the themes reviewed earlier, these six papers are less focused on laws and legal practices. Although they do refer to related domestic and international legal regulations, their primary focus is on scientific and technical matters. Nevertheless, their diagnoses of the maritime transport and ship-building industries conclude that the lack of systematic state planning and coordination are the main cause of the industries' current problems. Their prescriptions for the problems call for adequate legal and institutional devices established by the state.

The eighth and ninth themes involve marine scientific research and marine education and culture. There are seven papers that discuss these two themes. Four deal with the issues of marine science research on Taiwan, emphasising the importance of disseminating up-to-date scientific knowledge to the public, training marine biologists, supporting marine science to manage the threat of climate change, utilising advanced technology to generate green energy from the ocean and making more effort to conserve marine resources. The other three papers deal with the issue of Taiwan's marine education and culture. Modern Taiwan's religious practices, literary works, art performances and popular festivals related to the ocean are introduced.

The deficiencies of marine education are also reviewed. Although the ROC's marine education policy and actions for students are mentioned, the papers are concerned with the state's policy interests, not implementation; each presents its author's position on what the state should add to marine education policy development. They propose what marine education in Taiwan should be but do not explore the political issues related to the education.

As I suggested at the beginning of this sub-section, this journal's papers are sector-based, and all are legal and institutional; not only do they focus on legal and institutional analyses, but also their policy prescriptions involve amending existing laws, regulations and procedures and establishing new institutions and organisations. Additionally, their discussions of marine policies on Taiwan are more with regard to marine policy requirements and less to its implementation. The politics embedded in these processes is not examined. Furthermore, all of the papers are sector-based because each concentrate on a single marine policy issue of interest to the author without considering the structural context in which the particular marine policies are situated. It is worth repeating that my discussion of the social constructions of the ocean by the power sources in modern Taiwan society provides a way to comprehend the structural context.

Wen-yan Chiau (2017), an important member of the Institute of Marine Affairs and Policy, contributed a paper to the journal in which he calls for establishing a governmental organisation that can coordinate all state policies, decisions and actions related to maritime and ocean affairs. Currently, the sector-based attribute exists not only in the study but also in the actual practice of marine policy. This characteristic is evident even in the latest version of the ROC's 2020 National Ocean Policy White Paper (Council of Marine Affairs 2020). The white paper appears to be comprehensive in terms of the scope of maritime and ocean affairs that it seeks to cover. However, it effectively aggregates the leading issues and the state's proclaimed solutions.

This approach to formulating a national ocean policy is not problematic in and of itself; the United Kingdom, for example, has adopted a similar approach (Smith *et al.* 2015). Nonetheless, Chiau (2017) criticises this sector-based trend in the state's implementation of marine policy. He urges the enhancement of the state's coordination of governing Taiwan's maritime and ocean affairs by establishing a central governmental department (or committee) responsible for coordination. This is an institutional approach. In addition, Chiau (2017) considers the Coastal Zone Management Law insufficient for the state's effective governance of the ocean. He suggests that the government pass legislation to govern Taiwan's marine territory, including Jinmen and Mazu, to supplement the ROC's current legal structure regulating its marine territory.

In this sub-section, I have argued that the academic studies of Taiwan's marine policies are fragmented and disconnected. Each scholar has focused on a single issue in the field of maritime and ocean governance, taking a legal-institutional approach to analysis and prescription. An exception is

Chiau's (2017) paper, which calls for stronger, centralised coordination by the state of all maritime and ocean affairs in the society of modern Taiwan. His advocacy for building a comprehensive legal system of ocean governance is in line with the ROC's 2020 ocean white paper. However, the white paper and Chiau's (2017) proposal still fall within the legal-institutional paradigm; the political dimension of marine policy is not their main concern.

A Trend in Marine Policy Research and Advocacy: Integrated Ocean Policies

Chiau's (2017) call for the coordination of maritime and ocean affairs by the state echo the trend in marine policy research and advocacy in other countries. This sub-section briefly discusses how integrated ocean policies are examined. The discussion is supported by several publications. My discussion of integrated ocean policies by no means downgrades the importance of sector-based marine policies. If sector-based marine policy can be seen as one extreme of a continuum, then the integrated ocean policy can be viewed as the other.

The previous sub-section that analysed the articles published on Taiwan's marine policies concluded that the focus of the studies was primarily legal-institutional; an integrated ocean policy was not widely studied or promoted in the society of Taiwan. Therefore, how integrated ocean policy is analysed outside Taiwan is the focus of this sub-section.

What constitutes an integrated ocean policy can be defined first. Zacharias (2014) provides a classification of integrated ocean policies. The first type is integrated policy approaches, which 'are attempts to realise efficiencies from aligning current sector-based policies' (Zacharias 2014: 279). According to this definition, Chiau's (2017) paper and the ROC's ocean white paper are integrated policy approaches. The written works discussed later also belong to this type.

The second is integrated ocean management, which 'uses specific approaches and tools to assist with marine decision-making' (Zacharias 2014: 279). Integrated ocean management includes various methods, such as coastal zone management, ecosystem approaches to management, marine protected areas, marine spatial planning, cumulative effects and private ocean rights. In Taiwan's academic community, there have been only occasional mentions of topics in this field (for example, marine spatial planning and marine protected areas), but they have not usually been discussed or promoted in an integrated manner. Most importantly, potential political aspects of these ocean management issues are not considered. Because I only intend to illustrate that the studies and advocacies of marine policies generally follow a legal-institutional approach in the existing literature, Zacharias' (2014) introduction of integrated ocean management and its constituents is not discussed.

Early in 1980, Arild Underdal (1980), a reputable Norwegian political scientist, had already analysed an integrated marine policy, why it is needed and how it can be developed. The integration of marine policies involves unifying various related sector-based policies, which must satisfy the requirements of comprehensiveness, aggregation and consistency.

Comprehensiveness refers to the coordination and interdependence in the dimensions of time, space, actors and issues. Adopting a long-term vision in marine policies and evaluating the merits of the policies according to results constitutes the dimension of time in the comprehensiveness requirement. Controlling the consequences of policies within certain geographical or other types of spatial classifications is the space dimension. Evaluating the possibilities of all options for every concerned actor of a particular marine policy is the actor dimension. In the case of the issue dimension, comprehensiveness can be demonstrated in light of the extent of issue interdependence.

The aggregation of marine policies denotes 'the extent to which policy alternatives are evaluated from an "overall" perspective rather than from the perspective of each actor, sector etc.' (Underdal 1980: 161). The consistency requirement refers to a policy 'that is in harmony with itself – one whose different components accord with each other' (Underdal 1980: 161).

As to why marine policies should be integrated, Underdal (1980) convincingly suggests that integrated marine policies are needed because they can, and should, internalise the externalities generated by each sector-based, fragmented marine policy, which can result in collective inefficiency.

As to how integrated marine policies can be developed and implemented, Underdal's (1980) approaches are, remarkably, institutional. He points out that a direct approach 'seeks to achieve integration through defining policy goals and guidelines to be followed by all government agencies involved' (Underdal 1980: 166). By contrast, indirect approaches to the integration of marine policies can be sub-classified into intellectual and institutional measures (Underdal 1980).

Intellectual measures attempt to obtain marine policy integration 'through initiating research, training and socialization aimed at developing a more comprehensive and holistic perspective on the part of decision makers' (Underdal 1980: 167). Institutional measures include moving the issue-management level upward to the central governmental level, changing marine policymaking procedures, redistributing resources and formal authority amongst organisations and establishing new institutions for integrated marine policies (Underdal 1980).

Jean-Pierre Levy (1988), a former director and deputy to the Special Representative of the Secretary General for the Law of the Sea of the Office for Ocean Affairs and the Law of the Sea of the United Nations, has examined academic research trends on marine policies in developing countries after UNCLOS was established in 1982. He indicates that the concept of a national integrated marine policy was broadly promoted thereafter. Coastal states, especially developing ones, now faced a challenge of 'adopting the proper legal and institutional framework to establish a high-level policy in tune with their overall developmental objectives' (Levy 1988: 327). He also convincingly argues that the primary difficulty could be found 'in the ways and means to establish such a policy' (Levy 1988: 327).

The prescription that Levy (1988) identifies is, nonetheless, also legal and institutional. In what he calls the second-best, more realistic situation, the state would collect the necessary information to formulate an institutional base of knowledge about the marine conditions and the pre-existing uses of the ocean. Then, the state would establish an institutional planning mechanism to incorporate the input of all sectoral information. Subsequently, the state would build the institutional means to generate 'recommendations, alternatives, and evaluations, based on cost-benefit analysis, for decision-making' (Levy 1988: 328). Additionally, the policy decisions of the state would need to be translated into applicable laws and regulations through legislation and institutional mechanisms of policy implementation (Levy 1988).

Levy (1988) emphasises the importance of incorporating the uses of the ocean and its resources within the state's national economic development plan. This would require that policymakers pull the issues of marine governance back to the central stage and make efforts to establish procedures to resolve conflicts between the various, interdependent sectors of ocean utilisation. Ultimately, the above institutional prescription would depend on whether the state could foster a sea-oriented national culture that fully appreciated 'the importance and potential of the resources of their maritime environment' (Levy 1988: 329). Therefore, it would be reasonable to expect that an effective integrated marine policy could 'set out priorities and be translated through appropriate administrative mechanisms into specific projects, preferably complementary, but at least not in conflict with each other' (Levy 1988: 330).

D. Cameron Watt (1990), a previous Chair of the Sea Use Programme at the London School of Economics and Political Science, offers an insightful investigation into the impact of the central governmental executive-legislative relationships on how integrated marine policies can be developed. For Watt (1990), depending on the system of government, for example, presidential, cabinet-parliamentary, semi-presidential and politburo systems, different marine policy coordination and integration can be established.

Watt (1990) suggests that in a presidential system, because of the institution of checks and balances, the executive power for marine policies is solely conferred on the presidency; individual ministers exercise their power within the sphere of their own departments without much integration amongst themselves if the president does not coordinate. Watt (1990) suggests that in a semi-presidential system, interdepartmental rivalries or conflicts have to be resolved and coordinated by organisations attached to the president's or prime minister's office. The number of these organisations is usually limited, and their coordinating capacity is often concentrated on critical issues.

Watt (1990) argues that in a cabinet-parliamentary system, departmental rivalries can be coordinated in cabinet meetings. Marine policy integration is relatively easier to obtain 'at cabinet committee level between ministers or between officials nominated to represent their departments' (Watt 1990: 302). A possible obstacle to integration is the lack of recognition of the need for coordination (Watt 1990). Additionally, Watt (1990) argues that the integration of marine policies in a politburo system is similar to that of a cabinet-parliamentary system. However, the difficulty of recognising the need for coordination is aggravated in a politburo system. Moreover, in the politburo system, there may be 'the absence of an overall review system of comparative costs and the lack of responsibility of the planning departments for the execution of the planning targets' (Watt 1990: 302).

It can be argued that the legal and institutional approach to integrated marine policy studies and advocacies culminated in the publication of the *Routledge Handbook of National and Regional Ocean Policies*, edited by Biliana Cicin-Sain, David L. VanderZwaag and Miriam C. Balgos (2015). This handbook discusses national and regional integrated marine policies of 15 states and 4 geopolitical areas. The studies of the marine policies in these cases utilise a common analytical framework devised by the editors that considers the following themes: (a) The basic information of oceans and coasts in the territories of states and areas; (b) the momentum behind the initiation of marine policymaking; (c) the guiding principles derived from a range of international agreements and conventions, on which the marine policies are based; (d) the institutional arrangements and procedures for the initiations of marine policymaking; (e) the legislation for and legal frameworks of the marine policies; (f) the authority of organisations at various levels of the states' internal structures for marine policy implementation; (g) the organisations and institutions established for implementing international agreements and conventions domestically or within a region; (h) the organisations and institutions for implementing, financing, monitoring and evaluating marine policies and (i) the recommendations and advocacies for the states' and regions' ocean policies.

With each of the earlier themes, Balgos, Cicin-Sain and VanderZwaag (2015) provide a set of research questions. I do not consider it necessary to present them here. There are 27 sets of questions in their analytical framework, of which only 4 are explicitly about issues of actors and have the potential to develop into micro political analyses of marine policies (which I return to later). The other questions are all presented in a legal-institutional manner, which can be illustrated as follows.

For example, a question of theme (c) asks what international guiding principles had been adopted in a national integrated ocean policy. A question of theme (d) attempts to identify what institutional procedures are followed to develop an integrated ocean policy. A question of theme (g) inquires if there is any scientific research and technological development programmes to support a state's or a region's integrated marine policy (Balgos, Cicin-Sain and VanderZwaag 2015).

It must be clarified that, although I have suggested that Balgos, Cicin-Sain and VanderZwaag's (2015) analytical framework, as well as other papers and research findings discussed in this and the previous subsections, are legal-institutional in nature, I by no means downgrade the importance and usefulness of these works. If research on state-society relations can be considered as macro-level studies, as Mann (1994) convincingly argues, legal and institutional approaches to policy research and advocacies can be considered meso-level. As argued thus far, the existing literature on marine policy and the relevant analyses of maritime and ocean affairs generally concentrate on the meso-level.

As indicated by Leftwich (2004a; and also mentioned in Chapter 1 of this monograph), employing various definitions, theories, paradigms and approaches of a certain social phenomenon can lead to very different research results and conclusions. I, therefore, classify the main body of this monograph as a macro-level academic exploration, whereas the analytical framework I present below, and any other future research utilising the framework, is classified into the category of micro-level political analyses. It can thus be argued that this monograph and the framework can fill an academic gap in the existing literature.

A Flexible, Micro-level Framework for the Politics of Marine Policy Development and Implementation

Politics is the core of this micro-level analytical framework. To use this framework requires viewing politics as a process, not an activity that has taken place in a specific arena (Leftwich 2004a; 2004b). Politics is not restricted to state organisations. As explicated earlier and in Chapter 1, politics as interactions between and amongst actors, in which power is exercised to produce, distribute and use resources, takes place not only between and amongst societal powerholders but also within the state and between the state and society.

Additionally, as indicated earlier and in Chapter 2, the state is a member of society within the state's territory (and in the case of the ROC, it is one of the most powerful actors in the society of Taiwan after 1949). The state's policies that set the courses of action regarding the governance of maritime and ocean affairs will inevitably involve political interactions within the state and between the state and society. Moreover, related societal powerholders will undertake political actions and interact politically amongst themselves in response to the state's marine policies.

Here, Mann's (2012b) concepts of 'despotic power' and 'infrastructural power' of the state can help build this framework. According to Mann (2012b; and also as explicated in Chapter 1), the state's despotic power refers to the state's autonomy to make decisions and take actions without negotiating with societal powerholders. The state's infrastructural power denotes the state's capacity to properly implement its policies and decisions, once the policies are developed and decisions are made, no matter how despotic the state is.

There are at least five fields that can be explored by political science. Political analyses of marine policy development and implementation can, first, examine the political processes of marine policy development at the central governmental level: For example, how a prime minister interacts with parliament and exercises power to legalise and finance a marine policy is worth studying.

A second field is the central-local government political interactions regarding the development of marine policies: For example, how the central government's plan to designate a portion of a sea area of a local county can be made with or without the latter's opposition is a political process worth evaluating.

Third, what roles do societal powerholders in the ideological, military and economic spheres related to maritime and ocean affairs play in the development of the state's marine policies? This question is insinuated by the concept of despotic power. For example, do marine merchants and naval forces play any role in the development of particular marine policies? This type of inquiry is different from questions inspired by the concept of infrastructural power.

Fourth, thanks to the term 'infrastructural power', the political interactions between the state and related societal powerholders for implementing the state's marine policies can be studied. The implementation of a particular marine policy can affect some ideological, military and economic elites. Thus, how the state implements policy requires, at least, the neutralisation of the oppositions from societal powerholders. How the state fosters support from the societal agents is also worth examining. Furthermore, opposition to the implementation of certain marine policies may originate within the state, at the central or local levels. Likewise, support for the policy also needs to be examined. Additionally, the state institutions (in the sense of neo-institutionalism) and those between the state and elites required to implement policy effectively and efficiently if possible, can be explored.

Fifth, due to the state's attributes of territorialised, centralised regulation and coordination of political activities and relationships (Mann 2012a), the state must deal with geopolitics. Thus, a state's efforts to implement marine policy, for example developing its sea power, will be both affected by and will affect the state's geopolitical relations with other states. It is also important to analyse whether or not the state and others have institutionalised their interactions regarding particular marine policies. For example, the international influence on and from, China's actual implementation of the twenty-first century Maritime Silk Road Initiative is effectively within the parameters of research on the politics of marine policy implementation.

Readers of this monograph now may wonder what the difference is between my framework and other works in the existing literature. At first glance, it seems that they all take institutional approaches. However, I suggest that my framework is at the micro-level of political analysis, focusing on political actions and the interactions of state leaders, societal elites and their coalitions in marine policy development and implementation. These political actors are critical for institution-building, policy development and implementation.

Turning back to Mann's (2012b) concept of infrastructural power, when a state has greater infrastructural power, policy (marine) implementation is easier and reaching policy objectives is more likely. In order to better understand what is meant by high infrastructural power, Weiss (2006) illustrates infrastructural power with the concepts of penetrative capacity, extractive capacity and coordinating capacity.

Penetrative capacity refers to how deeply the state's tentacles can extend into society. For example, in pre-modern China, the traditional, dynastic states could only officially administer affairs at the county level, whereas the CCP could control people through the party organisation down to the level of natural villages (Hou 2012). Extractive capacity refers to how many resources (often in the form of taxes and compulsory services) the state is capable of taking from society. For example, the CCP could mobilise 2.25 million peasants for logistics in the Huaihai Campaign in 1948 (Hou 2012: 188), whereas the KMT's military forces could only rely on pressgangs.

Indeed, as Weiss and Hobson (1995) convincingly suggest, 'the more institutionalized and co-operative state-society relations are, the stronger the infrastructural power the state enjoys' (Hou 2012: 33; see also Evans (1995) for a similar argument). The state's coordinating capacity is the highest form of infrastructural power; its coordinating capacity refers to how much co-operation and synergy the state can generate with its partners in society to achieve its goals, although the goals may also originate from outside the state (Weiss 2006). For example, the ROC's promotion of industrial capitalism and the market in the society of Taiwan after 1949 is a demonstration of coordinating capacity.

Higher state infrastructural power means a greater chance of successful implementation of marine policies. Notably, institutionalisation and state-society co-operation can be seen as two approaches to high infrastructural power, especially the state's coordinating capacity. Moreover, institutionalising 'the interactions between the state and societal forces for the implementation of the state's policies presumes that the state has obtained a certain level of conscious submission or co-operation of the involved forces' (Hou 2020: 403).

As pointed out by the academic literature on marine policies in Taiwan and in other important works in English on marine policies discussed above, institutions are critical. That is, appropriate institutions and the institutionalisation of state-society co-operation are important for increasing the state's infrastructural power (as measured by the state's penetrative, extractive and coordinating capacities) to implement marine policy. 'Institutions and institutionalisations, however, by no means can be taken for granted' (Hou 2020: 403).

As Leftwich and Hogg (2008) identify, the desired institutions are created by the relevant, involved actors, who are often elite. It is usually the 'elites and coalitions that determine the shape and trajectory of institutional development' (Leftwich and Hogg 2008: 4). Thus, it is useful for political analysts of marine policies to bring people back into the analyses. Leftwich and Hogg (2007) clarify what they mean by leaders, elites and coalitions. Leaders in all fields constitute elites. Elites denote 'those small groups of people, seldom more than 3 per cent of any given population or unit of analysis ... in formal or informal positions of authority and power who take or influence key economic, political, social, and administrative decisions' (Leftwich and Hogg 2007: 4).

The successful development and implementation of the state's marine policies, the institutionalisation of state-societal co-operation among powerholders, political stability, economic growth, societal harmony and any other social progress, in general requires 'co-operation, trust and 'synergy' between different interests, groups and organisations' (Leftwich and Hogg 2007: 5). Moreover, for Leftwich and Hogg (2007; 2008), it is insufficient to rely only on the state at the centre to achieve these goals because the state's infrastructural power is low. Therefore, coalitions, defined as 'individuals, groups, or organizations that come together to achieve social, political and economic goals that they would not be able to achieve on their own' (Leftwich 2012: 5, original emphasis), formed by determined state leaders and societal elites, with the participation of and support from other concerned elites, do matter in the process of policy (including marine policy) development and implementation (Leftwich 2012).

Additionally, the political analysis of elite coalitions is concerned not only with politics that occurs 'during the formation and within the operation of a coalition, but also when the coalition exerts its joint power to overcome common obstacles and oppositions' (Hou 2020: 403). This point is particularly important for understanding marine policy development and implementation. The analysis of the interactions between leaders and elites for the establishment of coalitions and suitable institutions to develop and implement certain marine policies is what I call the micro-level analysis of the politics of marine policies.

The presentation of the framework for the micro-level political analysis does not reduce the importance and usefulness of legal and institutional studies and advocacies of marine policies discussed earlier. Instead, the micro-level political analysis framework is built on the foundation laid by the existing literature. It is worth repeating that the major difference is that the framework now incorporates another level and examines the political interactions amongst relevant leaders and elites, both within and outside coalitions, regarding the development and implementation of marine policies. Theories suggest that suitable institutions pursuing economic efficiency and sound laws and regulations to achieve legal justice cannot be taken for granted; they must be established, maintained and cultivated by relevant state leaders and societal elites capable of forming coalitions powerful enough to overcome oppositions, incorporate assistance, generate synergies of co-operation and coordination and develop the corresponding state infrastructural power for the policies.

The micro-level political analytical framework, moreover, can be flexibly utilised to comprehensively explore the process of marine policy development and implementation. With the assistance of despotic and infrastructural power, the five possible fields of political study of marine policy mentioned above can be investigated separately or together. In addition, various topics of maritime and ocean affairs dealt with by state policies can be the subject of micro-level analysis. The framework can also be applied to national or regional integrated marine policy and to other sector-based marine policies.

Furthermore, not only can the framework for the micro-politics of marine policies be combined with legal, institutional research and advocacies at the meso-level, but it can also be connected with the main theme of this monograph: The social constructions of the ocean by the four most important types of sources of power in modern Taiwan. As suggested at the beginning of this section, according to Steinberg's (2001) theory of territorial political economy, the social constructions of the ocean, the material organisation of society for ocean utilisation and the geographical conditions of the ocean are mutually affected. When considering the politics of marine policies, the influence of societal organisations and natural marine situations require the appropriate evaluations.

As Steinberg (2001) identifies, the use, regulation and representation of the ocean shape each other, and their interactions have effects on the construction of the ocean. Thus, the examination of marine policies through the micro-level political framework of leaders, elites and coalitions demands incorporating considerations of their uses, regulations and representations of the ocean. It can now be argued that this monograph's main discussion serves to establish a contextual structure for the politics of marine policy development and implementation and that the micro-, meso- and macro-level analyses are connected.

Conclusion

The ROC's Ocean Basic Act (which proclaims that the ROC should build 'a high-quality marine country' as announced in Article 1) was promulgated in 2019, and its *National Ocean Policy White Paper* was published in 2020. The state, under the presidency of CAI Yingwen (also known as TSAI Ing-wen) has striven to implement a series of marine policies, including an offshore wind power generation policy. Nonetheless, the experiences of the ROC's difficulties in developing and implementing marine policies (both sector-based and nationally aggregated) remain to be studied, and prospective actions remain to be prescribed.

Chiau (2019) suggests many problems that have emerged from the offshore wind power generation policy, including repercussions for the oceanic ecosystem, the contradictions with location principles, superficial environmental impact assessments, the difficulty of inter-ministerial communications, the difficulty of policy coordination, confrontations for fishery compensation, the lack of basic research and survey, radar shielding concerns, the challenge of underwater heritage protection and conflicting laws and regulations.

Chiau (2019), as does the existing literature discussed earlier, again provides an institutional prescription for the problems and urges the state to conduct proper marine spatial planning. I do not need to summarise his solutions, nor do I endorse or criticise his viewpoints here. I mention this to show that the development and implementation of the ROC's marine policies still remain to be examined and evaluated.

In the second half of this chapter, I have argued that detailed accounts are needed of the political interactions undertaken by relevant state leaders and societal elites and of the political processes of coalition formulations to establish appropriate laws, institutions and state-society co-operation for developing and implementing marine policies.

Additionally, the structural-contextual factors that influence leaders, elites and their coalitions when taking policy actions to achieve their marine goals must be taken into account by analyses. The first half of this chapter summarises the conceptual tools for understanding these factors and recapitulates the social constructions of the ocean of the most important power sources in Taiwan between 1949 and 2016. As promised in Chapter 1, I have answered a fundamental question of marine policy research in Taiwan and established the basis for future studies on the politics of governing Taiwan society's maritime and ocean affairs.

Notes

1 My preliminary attempt to introduce the notion of elite coalitions and politics into the study of sea power can be seen in my paper, see Hou (2020).

2 In this chapter, the term 'institutional' follows the tradition of neo-institutionalism, which has been discussed in Chapter 1 of this monograph.

References

Balgos, Miriam C., Biliana Cicin-Sain, and David L. VanderZwaag (2015) 'A Comparative Analysis of Ocean Policies in Fifteen Nations and Four Regions', in Biliana Cicin-Sain, David L. VanderZwaag and Miriam C. Balgos (eds.), *Routledge Handbook of National and Regional Ocean Policies* (London; Routledge), 3–48.

Biddle, Stephen (2007) 'Strategy in War', *PS: Political Science and Politics*, 40(3), 461–466.

Bush, Richard (2016) 'Security Policy', in Gunter Schubert (ed.), *Routledge Handbook of Contemporary Taiwan* (Oxon; Routledge), 529–543.

Chiau, Wen-yan (2017) 'Maixiang Haiyang Guojia: Piaobuo de Hangcheng?' (Towards A Maritime Nation: A Wandering Journey?), *Haiyang Shiwu yu Zhengce Pinglun* (Marine Affairs and Policy Review), 5(1), 5–32.

Chiau, Wen-yan (2019) 'Cong Haiyang Zhengce Guandian Lun Taiwan Lian Fengchang zhi Fangzhan' (The Development of Offshore Wind Farms in Taiwan: A Marine Policy Perspective), *Guojia Fangzhan Yanjiu* (Journal of National Development Studies), 18(2), 55–124.

Council of Marine Affairs Advancement (2006) *Haiyang Zhengce Baipishu* (Ocean Policy White Paper) (Taipei; Research, Development and Evaluation Commission, Executive Yuan).

Council of Marine Affairs (2020) Guojia Haiyang Zhengce Baipishu (National Ocean Policy White Paper) (Kaohsiung; Council of Marine Affairs).
Giddens, Anthony (1987) *The Nation-State and Violence* (Berkeley; University of California Press).
Evans, Peter (1995) *Embedded Autonomy: States and Industrial Transformation* (Princeton; Princeton University Press).
Hou, Kuang-hao (2012) *State Domination in China: An Examination of the Applicability of the IMEP Model for the Analysis of Chinese Politics in the 20th Century* (Heidelberg, Australia; Heidelberg Press).
Hou, Kuang-hao (2014) 'The Meaning of the Ocean for the State in Contemporary Taiwan since 1949: Understanding State Perception of the Ocean through a Historical Approach', *Australian Journal of Maritime and Ocean Affairs*, 6(4), 191–206.
Hou, Kuang-hao (2020) 'An Eccentric Analysis of Political Logistics of Maritime Mastery: Establishing a Framework', *Journal of Historical Sociology*, 33(3), 389–409.
Mann, Michael (1994) 'In Praise of Macro-Sociology: A Reply to Goldthorpe', *The British Journal of Sociology*, 45 (1) 37–54.
Mann, Michael (2012a) *The Sources of Social Power, Volume 1: A History of Power from the Beginning to AD 1760*, 2nd edition (Cambridge; Cambridge University Press).
Mann, Michael (2012b) *The Sources of Social Power, Volume 2: The Rise of Classes and Nation-states, 1760–1914*, new edition (Cambridge; Cambridge University Press).
Marx, Karl (1996) *Marx: Later Political Writings*, edited and translated by T. Carver (Cambridge; Cambridge University Press).
Migdal, Joel (2001) *States in Society: Studying How States and Societies Transform and Constitute One Another* (Cambridge; Cambridge University Press).
Murray, William (2008) 'Revisiting Taiwan's Defense Strategy', *Naval War College Review*, 61(3), article 3, accessible at: https://digital-commons.usnwc.edu/nwc-review/vol61/iss3/3, accessed on 24 February 2020.
Lee, Kuo-tien (2007) 'Taiwan Yan Ji Hai Yuye zhi Tantao' (Discussing Taiwan's Near-shore and Near-sea Fisheries), in Shinwen Weng (ed.), *Taiwan Yuye Zhengce Yanjiu* (A Study of Taiwan's Fishery Policy) (Taipei; Chung Cheng Agriculture Science and Social Welfare Foundation), 64–83.
Leftwich, Adrian (2004a) 'Thinking Politically: On the politics of Politics', in Adrian Leftwich (ed.), *What is Politics? The Activity and its Study* (Cambridge; Polity Press), 1–22.
Leftwich, Adrian (2004b) 'The Political Approach to Human Behaviour: People, Resources and Power', in Adrian Leftwich (ed.), *What is Politics? The Activity and Its Study* (Cambridge; Polity Press), 100–118.
Leftwich, Adrian and Steve Hogg (2007) 'The Case for Leadership and the Primacy of Politics in Building Effective States, Institutions and Governance for Sustained Growth and Social Development for Sustained Growth and Social Development', Background Paper 01, *Developmental Leadership Program*, retrieved from http://www.dlprog.org/publications/background-papers, accessed on 20 March 2021.
Leftwich, Adrian and Steve Hogg (2008) 'The Politics of Institutional Indigenization: Leaders, Elites, and Coalitions in Building Appropriate and Legitimate Institutions for Sustainable Growth and Social Development', Background Paper 02, Developmental Leadership Program, retrieved from https://www.dlprog.org/publications/background-papers, accessed on 20 March 2021.
Leftwich, Adrian (2012) 'Introduction: Coalitions in the Politics of Development', in *A Research Report and Policy Workshop Report: Coalitions in the Politics of Development:*

Findings, Insights and Guidance from the DLP Coalitions Workshop, 15–16 February 2012, Sydney, Australia; Developmental Leadership Program, retrieved from https://www.dlprog.org/publications/background-papers, accessed on 11/April/2021.
Leng, Tse-kang (2016) 'Taiwan and Economic Globalization', in Gunter Schubert (ed.), *Routledge Handbook of Contemporary Taiwan* (Oxon; Routledge), 233–250.
Levy, Jean-Pierre (1988) 'Towards an Integrated Marine Policy in Developing Countries', *Marine Policy*, 12(4), 326–342.
Peters, B. Guy (1999) *Institutional Theory in Political Science: The 'New Institutionalism'* (London; Pinter).
Research, Development and Evaluation Commission (2001) *Haiyang Baipishu* (Ocean White Paper) (Taipei; Research, Development and Evaluation Commission, Executive Yuan).
Smith, Hance D., Iwan Ball, Rhoda C. Ballinger, Tim Stojanovic, and Tara Thrupp, (2015) 'Strategy for Marine Environment Conservation and Development in the United Kingdom', in Biliana Cicin-Sain, David L. VanderZwaag, and Miriam C. Balgos (eds.), *Routledge Handbook of National and Regional Ocean Policies* (London; Routledge), 186–235.
Steinberg, Philip E. (2001) *The Social Construction of the Ocean* (Cambridge; Cambridge University Press).
Underdal, Arild (1980) 'Integrated Marine Policy: What? Why? How?', *Marine Policy*, 4(3), 159–169.
Wade, Robert (2004) *Governing the Market: Economic Theory and the Role of Government in East Asian Industrialization*, 2nd edition (Princeton; Princeton University Press).
Watt, D. Cameron (1990) 'An Integrated Marine Policy', *Marine Policy*, 14(4), 299–304.
Weiss, Linda (2006) 'Infrastructural Power, Economic Transformation, and Globalization', in John A. Hall and Ralph Schroeder (eds.), *An Anatomy of Power: The Social Theory of Michael Mann* (Cambridge; Cambridge University Press) 167–186.
Weiss, Linda and John Hobson (1995) *States and Economic Development: A Comparative Historical Analysis* (Cambridge; Polity Press).
Zacharias, Mark (2014) *Marine Policy: An Introduction to Governance and International Law of the Oceans* (London; Routledge).

Index

Page numbers followed by n denote notes.

For Product Safety Concerns and Information please contact our EU representative GPSR@taylorandfrancis.com
Taylor & Francis Verlag GmbH, Kaufingerstraße 24, 80331 München, Germany

www.ingramcontent.com/pod-product-compliance
Ingram Content Group UK Ltd.
Pitfield, Milton Keynes, MK11 3LW, UK
UKHW022316250926
475702UK00046B/869
9781032270982